Solar Photovoltaics

N. D. Kaushika • Anuradha Mishra • Anil K. Rai

Solar Photovoltaics

Technology, System Design, Reliability and Viability

Springer

CP

N. D. Kaushika
Centre for Energy Studies
Indian Institute of Technology Delhi
Delhi, India

Anuradha Mishra
National Institute of Technology Calicut
Calicut, India

Anil K. Rai
Department of Electrical & Electronical Engineering
Ajay Kumar Garg Engineering College
Ghaziabad, Uttar Pradesh, India

Co-published by Springer International Publishing, Cham, Switzerland, with Capital Publishing Company, New Delhi, India.

Sold and distributed in North, Central and South America by Springer, 233 Spring Street, New York 10013, USA.

In all other countries, except SAARC countries—Afghanistan, Bangladesh, Bhutan, India, Maldives, Nepal, Pakistan and Sri Lanka—sold and distributed by Springer, Tiergartenstr. 15, 69121 Heidelberg, Germany.

In SAARC countries—Afghanistan, Bangladesh, Bhutan, India, Maldives, Nepal, Pakistan and Sri Lanka—printed book sold and distributed by Capital Publishing Company, 7/28, Mahaveer Street, Ansari Road, Daryaganj, New Delhi 110 002, India.

ISBN 978-3-319-72403-4 ISBN 978-3-319-72404-1 (eBook)
https://doi.org/10.1007/978-3-319-72404-1

Library of Congress Control Number: 2017964703

Printed on acid-free paper

This Springer imprint is published by the registered company Springer International Publishing AG part of Springer Nature.
The registered company address is: Gewerbestrasse 11, 6330 Cham, Switzerland

Preface

Decentralized electricity-generating units are effective means of energy conservation. However, being based on diesel generation, these offer only a short-term solution as the world's fossil fuel production has already started declining. In this regard, mature and sustainable decentralized electricity generation technologies such as wind power and solar photovoltaic, solar thermal electric and hybrid renewable power systems are being investigated around the world. Solar conversion technology based on photovoltaics has several positive attributes and is on track to become a leading competitor for next-generation power if system efficiency and cost-effective approaches could be addressed.

In recent years, extensive research based on a multitude of solar cell technologies and balance of system (BOS) components has exhibited the potential for significant cost reduction and efficiency as well as functionality amelioration to make solar photovoltaics a viable proposition. In this monograph, an overview of global research activities on solar photovoltaics is presented. In particular, Chap. 1 encompasses the introduction to solar photovoltaic power and power generation from other renewable resources; Chap. 2 deals with solar radiation measuring instruments and an estimation of average solar radiation on a tilted surface; Chap. 3 presents a review on the fundamentals of photovoltaic generation and formation of *p-n* junction in semiconductor materials; Chap. 4 deals with different solar cell materials and wafer-based fabrication technologies, while Chap. 5 exposes mathematical models of transport processes in solar cells; Chaps. 6 and 7 introduce the electrical characteristics of solar cells and power loss in array networks due to mismatch in solar cell characteristics; Chap. 8 displays BOS other than solar PV array and different types of electronic regulations; Chap. 9 presents the repertoires of space, terrestrial and cost-effective applications of solar PV systems, while Chap. 10 discusses a knowledge-based approach for the design of solar PV system and web access to solar PV design aid expert systems for mass media adaptation. The reliability considerations of solar PV systems and discussion on high-performance solar cells starting from first-generation to third-generation solar cells are presented in Chaps. 11 and 12, respectively, whereas Chap. 13 exposes the cost-benefit analysis of engineering systems/projects.

This monograph covers some of the materials required for solar photovoltaic courses that are being taught at engineering colleges and departments of some universities. It is intended to serve research students, scientists, engineers, policy experts and programme managers. Sincere thanks are due to Dr. R.B. Gangadhar, for the useful discussions during the preparation of the monograph.

Delhi, India N. D. Kaushika
Calicut, India Anuradha Mishra
Uttar Pradesh, India Anil K. Rai

Contents

About the Authors

N. D. Kaushika Ph.D. formerly professor at the Centre for Energy Studies, Indian Institute of Technology Delhi, and subsequently director of reputed engineering institutions in Delhi and the National Capital Region, is a specialist in renewable energy and environment. He has been a recipient of S.S. Bhatnagar Research Endowment Award (Hariom Ashram Prerit) for research in energy conservation in 1987. He has promulgated the establishment of several research centres in engineering colleges, authored four books and contributed to articles in several reputed journals as well as authored chapters in books by international publishers as Academic Press, USA and Elsevier.

Anuradha Mishra Ph.D. joined the Centre for Energy Studies, IIT Delhi, and obtained her Ph.D. in 2004 specializing in "modelling and simulation of solar power systems". Presently, she teaches both graduate and postgraduate courses at the National Institute of Technology Calicut (NITC) where she is also an independent researcher.

Anil K. Rai Ph.D. was awarded his Ph.D. from IIT Delhi in 2008. He was the coinvestigator in the Department of Science and Technology, Government of India that sponsored research projects on solar photovoltaics. Presently, Prof. Rai works at the Department of Electrical and Electronics Engineering at the Ajay Kumar Garg Engineering College affiliated to Dr. A.P.J. Abdul Kalam Technical University, Lucknow. He has published several articles in international journals. His areas of interest are modelling and simulation of renewable energy systems, system engineering and electrical power quality.

Chapter 1
Introduction to Solar Photovoltaic Power

1.1 Conventional Power Supplies

Electricity is the most convenient form of energy for industrial, agricultural, commercial and domestic activities; it is the key input for the economic and industrial development of a country. In recent years, countries all over the world have taken up industrialization, transport, communication as well as automation projects on large scale and to maintain reasonable growth they must have a sustainable electricity supply to achieve their developmental goals. However, the current state of affairs is quite dismal due to the following challenges.

1.1.1 Gap in Supply and Demand

The developmental activities have led to an ever increasing escalation in the gap between supply and demand for electricity. This is more pronounced in developing countries where there is a chronic shortage of grid electricity caused by large amount of generation and distribution losses. It may be illustrated from the facts and figures for Indian electricity supplies as: power deficit 2.6%, peak power deficit 3.2%, transmission and distribution losses 23% and plant load factor of 65%.

1.1.2 Fuel Security Concerns

The per capita consumption of electrical energy is a reliable index to connote the progress made by a nation (for example, for 2011 per capita consumption of electrical energy was 753 kWh for India, 4332.5 kWh for UK and 9538.8 kWh for

N. D. Kaushika et al., *Solar Photovoltaics*,
https://doi.org/10.1007/978-3-319-72404-1_1

USA). So if the standards of living in the developing world are to improve, manyfold increase in the electricity supplies is needed.

Presently major means of electrical generation include: hydro generation using falling water resources, fossil fuel generation using coal, oil gas and nuclear generation using fissionable nuclear materials. The Earth has fixed amount of non-replenishable resources of fossil fuels and nuclear materials and it has been estimated that the proven reserves (that is recoverable with today's technology) will be exhausted in about 50–200 years. Hydro power generation is restricted to geographically and geologically suited locations and availability of potential energy of water. The accelerated generation from fossil fuels is also not encouraging due to limited reserves of coal, oil and gas, more so, in view of oil crisis in recent years. All these resources are able to cater to world's increasing power requirements partially.

1.1.3 Environmental Concerns

The excessive use of fossil fuels (coal, oil and gas), nuclear fuel and hydro resources is not benign to the environment and ecological balance on earth. For example, USA has already developed around 50% of its hydro potential and further expansion is suffering a set back because of ecological considerations. The adverse effect of the excessive mining and use of fossil fuels are well documented. For example, coal mining generates:

1. Ridges and valleys which result in erosion of top soil.
2. Waste material forms huge refuge banks occupying hundreds and thousands of acres of land.

The combustion of coal at the power plant site adds fly ash, smoke, soot and gaseous oxides of sulphur, nitrogen and carbon to the atmospheric air. For each kW of electricity installation about 2.5 tonnes of coal is burnt annually and a plant of 200 MW capacity annually emits 16,000 tonnes of toxins and residues. Furthermore, the typical conversion efficiency of power plant is 40%; consequently, 60% of heat of combustion of coal goes to air and water environment.

The carbon dioxide and heat added to atmosphere have resulted in the increase of earth's planetary temperature which is often referred to as 'Global Warming'; it is threatening the world in terms of reduction of polar snow cover and raising of the equatorial ocean level to flood the low lying islands and coastal regions; oxides of nitrogen and sulphur are posing a threat to plants and vegetations kingdom through 'acid rains'.

Nuclear power generation is to be carried out under stringent safety regulations which involve cost. The limitations due to hazards of nuclear waste material and its disposal impose further restrictions on nuclear power generation.

1.1.4 Rising Prices

Hydroelectric, thermal and nuclear power generating stations have following prerequisites:

1. detailed site investigations for the design and construction work of generators, distribution centres etc.
2. survey of load requirement and
3. laying of transmission cables, towers etc.

The above activities continue for several years and the project is often fraught with several socio-economic problems before the generating stations start their operations and deliver power to the user. In cases, where the generating point is situated far away from the location where the power is to be used, the cost of transmission lines and their maintenance makes the cost of power prohibitive and remote areas suffer a setback in electrification.

In view of above challenges, non-conventional methods of power generations have received steadily increasing attention around the world to control

1. dependence on fuels with escalating costs; and
2. pollution resulting from the use of nuclear and fossil fuels.

Non-conventional methods of power generation include solar photovoltaics, solar thermal, bioconversion, small hydro, wind and ocean temperature gradients and currents, which are also known as renewable resources of energy. These resources are replenished almost as fast as they are used and they do not lead to environmental pollution. They utilize sunlight, solar heat, biological waste, winds and ocean currents which are otherwise unutilized.

1.2 Power Generation from Renewable Resources

In recent years, power generation from such renewable resources as solar, wind, biomass, small hydro (including minimicro hydel), and ocean gradient have been seriously counted upon to bridge the gap between the global demand and supply of power. Extensive efforts over the past several decades aimed at research, development, testing and demonstration of renewable power technologies have been made throughout the world. Total worldwide R&D and demonstration investment on emerging renewable power technologies has exceeded hundreds of billion dollars. Substantial base of field experience and knowledge concerning these technologies has been established. In what follows we present a brief outline of these technologies.

1.2.1 Small Hydro

Hydro power is obtained from the potential and kinetic energy of water flowing from a height. A hydro power project of capacity up to 25 MW is referred to as small hydro power plant. The hydro power potential of a site is dependent on the discharge and head of water as follows:

$$P = \eta \times Q \times g \times H$$

where P = power in kW, η = overall power generating system efficiency, Q = discharge (rate of flow) in m^3/s, g = acceleration due to gravity (9.8 m/s) and H = head (height) in metres.

Small hydro power projects are classified on capacity as follows:

Micro hydro: up to 100 kW
Mini hydro: 101–2000 kW (i.e. 2 MW)
Small hydro: above 2 MW up to 25 MW

Small hydro power projects are also classified on the basis of head as follows:

Ultra low head: below 3 m
Low head: above 3 and up to 40 m
Medium/high head: above 40 m

Development of the small hydro power plant involves:

1. civil work making reservoir; and
2. choice of a suitable turbine and generator. In the simplest system a suitable centrifugal pump may be selected to work as a turbine by matching its reverse characteristic and coupling it to an alternator. The centrifugal pumps offer several such advantages as low cost, easy installation and operation but can provide only one discharge for a given head and speed whereas hydraulic turbine has flow and therefore load control.

The cost of mini/micro hydel power plants is a sensitive function of the magnitude of head; higher the value of head lower the cost. Illustrative cost of small hydro power plant in India is Rs. 7–8.5 crores per MW.

1.2.2 Biomass and Waste

Biomass includes agro-waste and residues such as bagasse, rice husk, straw, cotton stalk, coconut shells, soya husk, de-oiled cakes, coffee waste, jute wastes, ground nut shells, saw dust etc. These are also new and renewable resource for power generation. Municipal and industrial wastes can also be useful for electricity generation, but are basically different forms of biomass.

The processes for conversion of agro waste and residues into useful form of fuels/ energy include:

1. thermo chemical processes (gasification, pyrolysis and liquefaction);
2. biochemical processes (anaerobic digestion and fermentation); and
3. direct conversion.

Over the years all these processes have been tried and tested and shown technically feasible, but they differ considerably in their efficiency of conversion and degree of commercialization.

For electricity generation, gasification and direct combustion are the serious alternatives. Gasification is a century-old technology. In this process the biomass (agro-waste or residue) is reacted with a limited amount of air which is insufficient for combustion. An inflammable gas known as producer gas or fuel gas is generated. The fuel gas is a mixture of carbon monoxide, methane, carbon dioxide and hydrogen. The gas, after cleaning and cooling, can be used as gaseous fuel in diesel engines with 70–80% cut in diesel consumption. During the World War II many producer gas powered vehicles were used in Europe and Australia. Gasified-based power units are now commercially available; however, there are very few working systems around the world though there have been several reports of the success of experimental systems. This appears to be due to the fact that gasifier units are too complex and messy to be viable for small scale (10–100 kW) utilization. Furthermore, the producer gas is a fuel of relatively low energy density; consequently the combustion engine exhibits excessive wear and tear and the system maintenance is expensive. Similarly the product of the biochemical process is the biogas which also is a relatively low calorific value fuel (5000 k calorie/m^3) as against the common fuels like natural gas having calorific value (8600 k calorie/m^3). In contradiction to this is the process of direct combustion to raise steam. It is the simplest process and has been used for about 200 years. All the biomass is combusted directly so the efficiency of conversion is superior to gasifiers. Well-designed agro-waste combustion systems are commercially available and they exhibit good reliability and operational characteristics; the resultant ash is recycled back to the ecosystem. The system is used for electricity generation and its waste heat may be utilized for such applications as crop drying.

1.2.3 Ocean Gradient

Oceans are the largest natural collectors of solar energy. The upper layers of tropical ocean are heated by the absorption of solar radiation; the water at depth of about 100–500 m is colder due to flow from polar regions. The absorption of solar energy takes place according to Lambert's law of absorption. It states that the absorption of light decays exponentially with distance into the material. The reduction in the intensity of a ray of original intensity along a length x is

$$I_{(x)} = I_0 e^{-Kx}$$

where I_0 = Intensity of radiation at the surface $x = 0$, I = Intensity of radiation at distance x below the surface, and K = absorption coefficient (unit L^{-1}). Its value is 0.05–0.5 depending upon type of water.

A typical variation of temperature with ocean depth is illustrated in Fig. 1.1 (Sukhatme 1984).

There are two types of energy which can be harnessed from ocean: thermal energy from ocean surface warm water and mechanical energy from the tides and waves. Ocean thermal energy conversion technology uses this temperature difference for obtaining power from sea. The largest reported temperature difference is about 20 °C and about 10 °C is lost across heat exchanger walls. The resultant temperature difference corresponds to 3.4% Carnot efficiency. Typical OTEC turbo generators may have an efficiency of about 60%. So the net obtainable efficiency will be about 2%. The operation of system auxiliaries and pumping of large quantities of water will also consume power; this may well reduce the net efficiency to a negative value. In fact it is for GW range that OTEC efficiency becomes positive. The poor thermodynamic efficiency of OTEC system is obviously due to the small temperature difference between source and sink.

A dam is used for allowing the water to pass through turbines to convert tidal energy into electricity. Tides are driven primarily by the gravitational pull of sun and moon on the earth, whereas the waves are driven primarily by the wind. Tide induced currents are the underwater currents utilized to drive the tidal turbines in a similar way as wind turbines are driven by the wind.

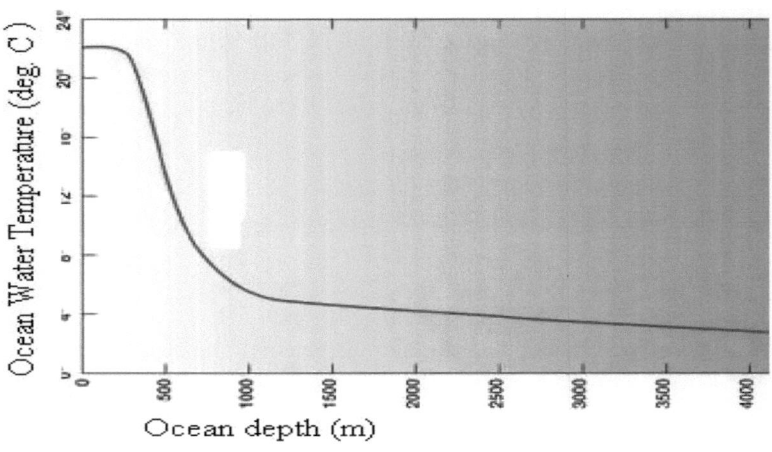

Fig. 1.1 A typical variation of temperature with ocean depth

1.2.4 Wind Power

Wind energy utilization is an ancient craft dating back thousands of years; it has been in vogue for various such applications as ship propulsion, milling grain, water pumping and more recently for generation of electricity. The basic principle involved in wind (air in motion) energy utilization is that the wind exerts pressure at an obstacle and is capable to do work; the rate of this work is called wind power. The global potential of wind power for electricity generation was estimated to be 432,419 MW; for India it was about 25,088 MW by the end of 2015.

Wind machines are classified on the basis of the direction of their axis of rotation. Two broad categories are: wind axis (or horizontal axis) machines and cross wind (or vertical axis) machines. The pumping wind mills have multi bladed rotors in a single wheel form generally known as fan-mill. The minimum wind speed required for a wind mill is 2 m/s and with a wind speed of 10 m/s it could develop power up to 30 kW; the cut out wind speed is 17 m/s. Wind electricity generators are propeller type with small number of blades (2 or 3) so that starting torque is low and they are fast running. Reliable wind generators ranging from 100 W–3 MW rated output are now commercially available. Some of the technological incorporations of 1980s in machines are fibre glass reinforced polyester blades, microprocessor controller and pitch regulations. The pitch regulation is rather a high tech feature; it allows the blades to pivot along their own longitudinal axis. At low wind velocity the pitch regulated blade offers large surface area against the wind and as the wind velocity increases the surface area decreases gradually. The resultant system enables optimum energy capture and has prolonged life span. It is cost effective in suitable wind regimes. Total estimated capacity of wind power generation in the world at the end of 2014 is 50 GW. The share held by Germany is 44.9%, USA has 4674 MW. Indian setup in Gujarat, Tamilnadu and Andhra Pradesh has total capacity of 1702 MW.

In the international market the capital costs of wind power system varies in the range of $1500–3000 /kW rated capacity; corresponding power costs are about 6–10 cent/kWh for grid connected systems. Generation costs of 5 cents/kWh are achievable in windy regions. Wind farms can be built for a total investment as low as US $1000 per installed kW. The cost of operation and maintenance for a wind power system is estimated as 1 cent/kWh. The cost of maintenance, operation and fuel for nuclear and coal based power plants are 1.9 and 2.3 cents per kWh respectively. In India grid connected power systems are being promulgated to an investment cost of Rs. 3.5 $\times$ 10^7/MW of rated output. The unit rating of individual generators range between 55 kW and 400 kW corresponding to rotor diameter of 15 m to 34 m. Each 55 kW unit saves 400 $\times$ 10^3 kWh of electricity per year and avoids the addition of 3–5 tonnes of carbon dioxide, 16–28 tonnes of slag and fly ash and 160–280 kg of particulate matter in atmosphere.

1.2.5 Solar Power Conversion Technologies

Solar radiant energy may be converted directly into electricity through processes like photovoltaic, thermoelectric and thermionic or indirectly through thermal conversion often referred to as thermodynamic conversion. The solar photovoltaics use a semiconductor device called solar cell for the conversion. It is in most advance stage of development as compared to other direct conversion methods. In the indirect method solar heat is produced in a collector or concentrator and imparted to a working fluid to increase its thermal energy which is subsequently converted into mechanical power through Rankine, Brayton or Stirling cycles or through specially designed devices.

1.2.5.1 Solar Thermal Electric

Solar radiant energy may be converted into electricity through thermal route often referred to as thermodynamic process. Solar thermal electric is the subject of intensive research and development activities in several developed countries, viz. the USA, Europe, Japan and Australia and in some developing countries such as India, Spain and Senegal etc. In thermodynamic process heat is collected from the sun rays in solar collector and is transferred in a heat exchanger to a working fluid which in turn activates a prime mover and alternator.

Several options in solar collectors, working fluid, prime movers, generators and storage units are now available. Based on these possibilities wide range of systems have been proposed and tested. Collector is the costliest component in a solar thermal electric system. The variations in collector technologies include solar chimney, solar ponds, flat plate collectors, evacuated tubular collectors, parabolic trough collectors, hemispheric bowl, parabolic dish and central receiver. The temperature range of heat generated by these collectors is illustrated in Table 1.1. Most advanced solar thermal power systems include parabolic dishes with high efficiency engines, central receiver power plants and solar boosted parabolic trough or solar ponds conventional thermal generators.

Table 1.1 Solar thermal technology and their temperature range

S. No.	Solar thermal power technologies	Temp. range (°C)
1.	Solar chimney	30–70
2.	Solar pond	50–80
3.	Flat plate collector	50–120
4.	Evacuated tubular collectors	90–200
5.	Parabolic trough	150–400
6.	Hemisphere bowl	200–500
7.	Paraboloidal dish	300–600
8.	Central Receiver	300–600

The parabolic trough has been explored by LUZ and subsequently by SOLEL. They have used solar parabolic trough collectors to preheat water prior to generating superheated steam in a conventional natural gas system. Given sufficient solar input the plant can operate at full rated power using solar energy alone. The parabolic trough solar technology is the most proven and lower cost large scale solar power technology available today. Field experiments on solar power tower concept have been carried out in Spain, Italy, Japan, France and USA. World's largest solar power plant of 10 MW capacity was constructed near Barstow (CA). It used Rankine cycle steam turbine operating at 500 °C. More recently a solar thermal test plant, based on central receiver concept, in Newcastle, Australia has generated super critical steam at a pressure of 23.5 MPa and 570 °C (Source: Dr. Alex Wonhas, CSIRO Energy Director).

The parabolic dish systems have the highest operating temperatures and hence higher conversion efficiencies. During 1980 to 2000 several investigations on design and materials of dish technology ware reported. Most notable development is that by reducing the weight of the structure, reduction in the system cost becomes possible (Kaneff 1983). A rationale with respect to economic design of a dish concentrator is to use the inherent strength of a paraboloidal shell involving a deep dish wherein no frame would be required which in turn would reduce the weight and cost. A deep dish will have a shorter focal length in comparison to a shallower dish (Kaneff 1983). The surface receiver can be used as a focal absorber when the rim angle is greater than 90°. A tradeoff between focal length, rim angle, nature of absorber and its life properties have been carried out by Kaushika (1986) to evolve a viable design involving deep dish of imperfect optics. Yet another viability approach is the dish-sterling system wherein heat to electricity conversion is carried by sterling engine mounted in the focal region of the dish. It is proved by gaseous hydrogen or helium, usual working fluids of sterling system.

The parabolic dish systems have an advantage over power tower of being modular. The estimated electricity generation costs of the three leading solar thermal technologies are as follows (Solar Directory 1989):

Technology	Installed costsU.S. $/kW	Overall generation U.S. cents/kWh
Parabolic dish	1400	6
Power tower	3000–2000	15–6
Parabolic trough	4000–2000	20–10

The facts and figures change over a period of time; the intention here is to give illustrative figures for comparative studies. The cost corresponds to systems larger than 5 MW and lower figure applies above 50 MW. The coal-based electricity costs are 4 cents/kWh. Thus in case of parabolic dishes the economics of scale is possible; lower power level and scaling up to larger size plants can be achieved in steps.

1.2.5.2 Solar Photovoltaics

Solar photovoltaic (PV) technology offers a convenient method for conversion of sunlight, available in abundance during a major part of the year, directly into electricity. The technique of conversion of sunlight directly into electricity by solar cells employing photovoltaic effect is environment friendly, easy to maintain and operate. The solar photovoltaic effect was first observed by Becquerel in 1839, when he directed sunlight on one of the electrode of an electrolytic cell. Later, Adams and Day in 1877 observed the effect in selenium. Amongst others the pioneering work of Lange (1930), Schottky (1930) and Grondhal (1933) led to the development of the photographic exposure meter and many other useful devices. Subsequently in 1954, Chaplin, Fuller and Pearson of Bell Telephone Laboratory fabricated diffused junction crystalline silicon cell with an efficiency of 6% and Reynold et al. (1954) made a similar device with cadmium sulphide. At the outset, single crystal silicon homojunction solar cell attracted considerable research and development for outer space as with the advent of the space age in 1950s, the need for a light and reliable power source for satellites was realized. The first solar powered satellite, the American Vanguard I, was launched early in 1958. Since then hundreds of satellites launched for scientific, military, meteorological, communications and other purposes have been powered by solar cells. Progressive improvements have raised the conversion efficiency of space solar cells to nearly 20%. At the outset, cost of solar cells was prohibitively high (US dollar 10 million per kW). Up to 1975, production of crystalline silicon solar cells, mainly for space programmes, averaged about 100 kW per year and the cost reduced to only 1 million US dollars per peak kW. Subsequently terrestrial applications of solar PV received attention; cost reduction has been the main objective. In this regard following aspects of solar photovoltaics have been investigated:

Solar Cells and Modules

The most common material used for solar cell is silicon which is available in abundance on earth. The technology of production of semiconductor grade crystal silicon is very well understood as this is used in fabrication of electronic components including diodes, transistors, integrated circuits and microchips. The manufacturing process of silicon PV modules comprises following steps:

1. Polysilicon production;
2. Single crystal ingot production;
3. Cutting ingots into wafers;
4. Cell production; and
5. Module assembly.

Based on the silicon wafer production technology, crystalline silicon cells are classified into three main categories:

- Monocrystalline referred to as single crystalline: these cells are circular in shape;
- Polycrystalline, referred to as multi-crystalline; and
- EFG ribbon silicon and silicon sheet-defined film growth.

These multicrystalline and EFG ribbon silicon cells are non-circular and do not involve waste of space in module formation.

Yet another classification based on the state and quality of the silicon material involves three types of solar cells that have been produced and marketed:

1. Mono-crystalline silicon solar cell;
2. Polycrystalline silicon solar cells; and
3. Amorphous silicon solar cell/Thin film cell.

Furthermore, dozens of PV materials and process technologies have been proposed and tested for the amelioration of efficiency and reduction of cost of solar cells.

Thin film solar cells based on amorphous Si (a-Si), Cu (InGa), Se2 (CIGS) and CdTe and nano/micro/poly-Si have received serious attention for their prospective high potential for cost reduction and quality improvement and are now receiving an increasing weight in the market. The laboratory cell efficiency is 25.6% for monocrystalline and 20.8% for multicrystalline wafer-based technology. The highest laboratory efficiency in thin film technology is 21% CdTe and 20.5% for CIGS solar cells (Fraunhofer ISE: 2016). The most common cells used in terrestrial applications as mentioned above are silicon cells. Silicon wafer-based PV technology is about 93% and thin film technologies are about 7% of the total annual production in 2015. The share of multi-crystalline technology is now 69% of total annual production.

Silicon might not be so able to meet long-term cost goals, since the thermodynamics fundamentally limit their conversion efficiency up to 31% or 41% depending on the concentration of incoming sunlight. This is known as Schockley-Quetsser efficiency limit.

Dye-cells and organic photovoltaics are also advancing and hold promise for affordable low cost technologies. Next-generation photovoltaics and nano-architectured solar cells may enable several advanced solar cells design concepts to be exploited (e.g. intermediate band and multiple exiton devices, hot carrier and up/down conversion devices, etc.) and predicted to be very high-efficiency approaches to solar energy conversion. However, the substantial theoretical and practical barriers in their implementation in useful devices for a commercial perspective are still to be removed. Theory and experiments directed at gaining a deeper understanding of the phenomena at atomic/molecular/nano- and meso-scale involved in the development of improved, emerging and new generation high efficiency/low cost solar cells are being currently researched.

Systems and Applications

The cells are connected in series/parallel to form modules which are used as a photovoltaic generator. Depending on the load requirement, these modules are connected to form arrays. Photovoltaic generators are modular in nature and can be set up quickly and easily as small arrays and increased in size as demand for power grows. So, solar PV generation is uniquely suited to remote areas and hilly regions as the sunshine including diffused radiation is a distributed source and it involves onsite generation cum utilization. For such areas, the transportation of fuel and laying of transmission lines becomes an uneconomical and unreliable proposition. Such a situation is prevalent in remote areas and hilly regions with scattered population where sunshine is abundant all round the year. Solar photovoltaic generators also provide power for several other terrestrial applications like telecommunications, water pumping, cathode protection of pipelines and other metallic structures, rural electrification, navigational aids, alarm systems, traffic warning lights and other consumer products like pocket calculators. The best applications correspond to distribution end of utility system or beyond the distribution system. The global cumulative solar PV installations reached to 242 GWp at the end of 2015 (Fraunhofer ISE: 2016) with a region wise status as follows:

S. No.	Country	Capacity of installations (%)
1	China	21
2	Germany	16
3	Italy	8
4	Rest of Europe	16
5	Japan	14
6	North America	13
7	Rest of World	12

Three configurations of SPV power plant system have been proposed and tested:

1. Non-tracking
2. Tracking
3. Tracking-concentrating systems

In the laboratory, high concentration multi-junction solar cells achieve an efficiency of up to 46% and with concentration technology module efficiency of up to 38.9% have been achieved (Fraunhofer ISE: 2016).

1.2.6 Comparative Study

Based upon our review of the technological activities, field experiences, cost trends and resource magnitudes, it may be concluded that wind power is closest to use in suitable windy regions. Solar appears to be a universal remedy for power generation.

The resource is best known, has appropriate magnitude, freely available and the technologies are widely applicable, non-polluting, environment friendly and deliver power directly or via thermal route. However the thermal route is cumbersome and troublesome. The direct conversion technology based on solar photovoltaics seems most promising. Its positive attributes may be summarized as follows:

1. Well Developed and Proven Technology
   Initially developed for space applications, the SPV technology has undergone exhaustive R&D, and has passed many stringent laboratory and field tests over a period of time spanning half a century.
2. Decentralized Generation of Power
   SPV solar plants can be setup and maintained in remote areas which are otherwise costly to connect with electrical transmission lines and roads.
3. Simple and Reliable
   Presently available solar PV power plants involve reliable electronic and electrical components that have been time tested.
4. Modular Configuration
   SPV power plants of all sizes can be developed using basic building blocks and can be easily installed in succession.
5. User and Eco-Friendly
   Solar power plants can be installed without safety problems and do not have any adverse effect on the environment and ecological balance known to this date.
6. Low Operation and Maintenance Costs
   Their capability for unattended operation and lack of need for water or other fluid for cooling confers in them low operation and maintenance costs and high degree of reliability.

1.3 Scope of the Monograph

In conclusion, solar conversion technology based on photovoltaics has several positive attributes and is on track to become a leading competitor for next generation green power if system efficiency and cost effective approaches could be addressed to make solar photovoltaic a commercial prospective. In recent years, extensive research based on micro and nano-solar cell technology has exhibited the potential for significant cost reduction and efficiency as well as functionality amelioration. This monograph presents an overview of global research activities in this direction which encompasses wafer-based solar cells materials and fabrication technologies, mathematical model of transport processes, mismatch losses in array network, BOS and electronic regulations, repertoires of applications, solar PV system design and reliability and prospective high performance solar cells. It considers mono-crystalline and multi-crystalline silicon-based solar cells and discusses the simulation model efficiencies as well as the practical efficiencies of thin film micro crystalline, nano crystalline, hybridized silicon-based solar cells, multi junction solar cells and thin film cadmium telluride (CdTe) cells. Furthermore, it points out that high efficiency of solar cell device reached so far is not followed by high system efficiency due to several drawbacks. In this regard it presents work on fault tolerant circuitry to combat mismatch losses in cell network and balance of the system. It also considers cost reduction options like building integrated photovoltaics and the concentrating systems.

Chapter 2
Solar Radiation Characteristics

2.1 The Solar Constant

Sun is a star, a vast spherical mass of hydrogen and helium in proportion 3:1 by weight; the proportion of other elements is less than 1% of total mass (1.9×10^{30} kg). The temperature in its central interior region is estimated to be very high ~2106 K; consequently huge amount of thermonuclear energy is generated in the central core by the process of nuclear fusion. Some of the basic characteristics of sun-earth system are illustrated in Fig. 2.1.

The radiant energy flowing out of the sun is 3.7×10^{26} Watts. This energy flow corresponds to a rate of depletion of solar mass as 4.1×10^6 tonnes per second or 1% of the total mass in billion years. The solar energy intercepted by the earth (S_E) may be calculated as follows:

$$S_E = \frac{3.7 \times 10^{26}}{4\pi R^2} \pi r^2$$

where R is the sun-earth distance $= 1.495 \times 10^{11}$ m and r is the earth's radius $= 6.35 \times 10^6$ m.

So we have $S_E = 173 \times 10^{15}$m.

The solar energy available outside the earth's atmosphere is called extra-terrestrial solar radiation. The amount of this energy that crosses in unit time through unit area placed perpendicular to the direction of energy flow is referred to as solar energy flux. At sun-earth distance this flux is essentially constant and is called the solar constant. Owing to the variation of sun-earth distance solar constant varies slightly with day of year as follows:

N. D. Kaushika et al., *Solar Photovoltaics*,
https://doi.org/10.1007/978-3-319-72404-1_2

Fig. 2.1 Sun-earth system

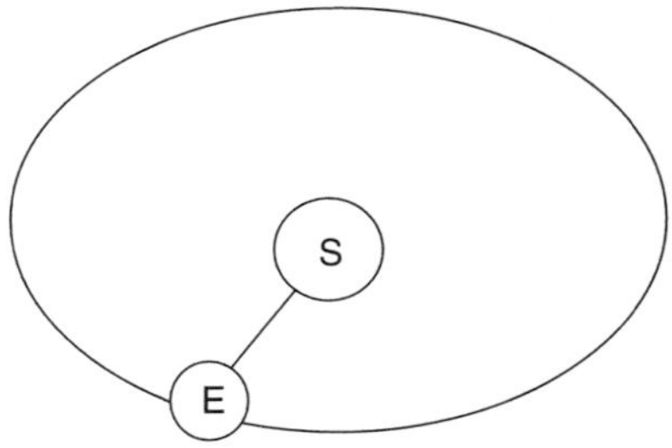

$$I_\alpha = I_{sc}\left[1 + 0.033\cos\left(\frac{360n}{365}\right)\right] \tag{2.1}$$

where I_α = extra-terrestrial radiation, I_{sc} = solar constant and n = day number of the year.

The expected value of solar constant is 1367 W/m^2 (Duffie and Beckman 1980).

2.2 Solar Spectral Distribution

Every hot body emits radiant energy which is characteristics of body temperature, size and nature of its surfaces. The spectrum of solar radiation too is characteristic of sun's temperature. However, since sun's energy is generated at its centre, which has a temperature of millions of degree, the central hot core terminates in a relatively colder surface called photosphere. For the purpose of solar spectrum and the corresponding temperature of sun, we always consider the temperature of photosphere. As viewed outside the earth's atmosphere, the spectrum of solar radiation is characteristic of that emitted from a black surface at a temperature of 5762°K. This implies that the solar spectral irradiance ranges between the wavelength region of 0.2–2.0 μm and has maximum invisible light region at 0.48 μm (Fig. 2.2). The brightness of solar disc also varies from its centre to edges. However, for engineering calculations, the disc is often assumed to be of uniform density.

2.3 Atmospheric Effects on Solar Radiation

Sun is a sphere of diameter 1.39×10^6 km situated at a distance of 1.50×10^8 km. It subtends an angle of 32 min at earth's surface. So sun's radiation beam outside the earth's atmosphere is almost parallel. During its passage through atmosphere, it is affected by the process of reflection, absorption (selective as well as continuous) and scattering due to clouds, particulate matter and gas molecules. All the processes are wavelength dependent and the magnitudes of effects depend on the thickness of the atmosphere penetrated. A measure of atmospheric thickness is air mass which is

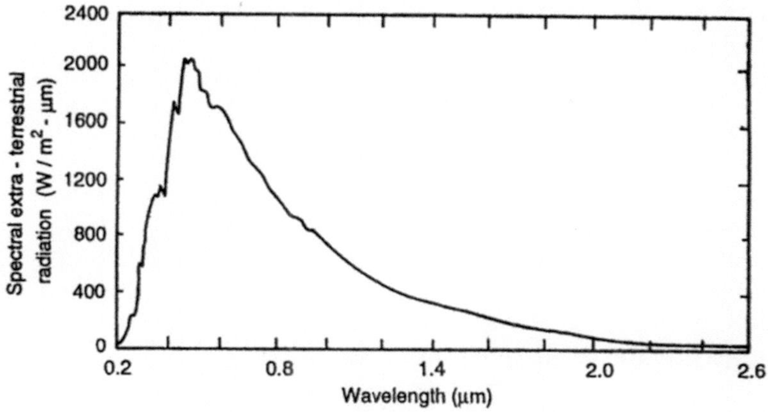

Fig. 2.2 Spectral distribution of Extra-terrestrial solar radiation

expressed by the ratio of the optical thickness of the atmosphere through which direct sun's rays pass to the optical thickness if the sun was at zenith (vertical incidence) and is given as follows:

$$m = (\cos \theta_z)^{-1}$$

where θ_z is zenith angle.

The net atmospheric effect is that radiation received at earth's surface consists of two components: (i) beam radiation—the radiation received from sun without change of direction and (ii) diffuse radiation—the radiation received from all directions. The sum of beam and diffuse radiation is referred to as total radiation or global radiation.

2.4 Solar Radiation Geometry

For the simulation and design of solar energy systems one is interested in the value of solar intensity on a surface having an arbitrary orientation; this in turn depends on the angle which the sun's rays make with the normal to the surface. This angle is referred to as incidence angle, θ for the planar surface. Solar incidence angle depends on the latitude of location (φ), day of the year characterized by the declination angle (δ), solar hour angle (ω), surface azimuthal angle (r_p), and slope of planar surface to horizontal (β). Several texts have dealt with the sun-earth geometry related to solar incidence angle. In what follows we present a discussion closely based on Boes (1981) and Sukhatme (1984).

Earth revolves round the sun in an elliptical orbit and this motion may be represented by sun's virtual motion in celestial sphere whose centre is centre of earth. The plane of the elliptical orbit is called the 'ecliptic plane'.

The angle between the sun-earth vector (which lies in elliptic plane) and earth's equatorial plane is called the solar declination angle, δ. It is maximum (23.45°) on June 21 and minimum (−23.45°) on December 21. It may be expressed as

$$\delta \text{ (in degrees)} = 23.45 \sin\left[\frac{360}{365.24}(n + 284)\right]$$

where n is day number of the year (1st January onwards).

The rotation of earth about its axis causes the day-night cycle and is the basis of solar hour angle which is an angular measure of time and is equivalent to 15° per hour. It is measured from solar noon (i.e. when the sun is on the local meridian) and is positive (negative) before (after) solar noon.

The latitude of site, φ, is the angle made by the projection of the line on the equatorial plane with the radial line joining the site to the centre of earth. It varies from −90° to +90° and measured as positive for northern hemisphere conventionally.

The sun's position on celestial sphere may be represented by solar azimuth angle, r_s (sun's angular distance from the south) and solar altitude angle, α (sun's angular distance from horizon). Similarly the orientation of fixed planar surface is represented by its azimuth angle, r_p (angle in the horizontal plane between line due south and the projection of normal to the surface on the horizontal plane) and its tilt angle with the horizontal, β. The azimuth angle is measured from the south, positive towards east. As an illustration $r_p = 0$ for south wall, +90° for east wall, −90° for west wall and 180° for north wall of a building. β is taken positive for south facing surfaces. All the above angles are illustrated in Fig. 2.3.

Following Boes (1981), solar incidence angle θ may be expressed as:

$$\cos\theta = \cos\alpha\cos\left(r_s - r_p\right)\sin\beta + \sin\alpha\cos\beta \tag{2.2}$$

where solar altitude angle α is given by

$$\sin\alpha = \cos\phi\cos\delta\cos\omega + \sin\phi\sin\delta$$

and the solar azimuth angle, r_s by:

$$r_s = \sin^{-1}\left(\frac{\cos\delta\sin\omega}{\cos\alpha}\right) \text{if } \cos\omega > \frac{\tan\delta}{\tan\phi}$$

and

$$r_s = 180 - \sin^{-1}\left(\frac{\cos\delta\sin\omega}{\cos\alpha}\right) \text{if } \cos\omega < \frac{\tan\delta}{\tan\phi}$$

Following simple cases are often of interest:

1. Vertical surface (viz. walls)

Fig. 2.3 Sun-earth angles

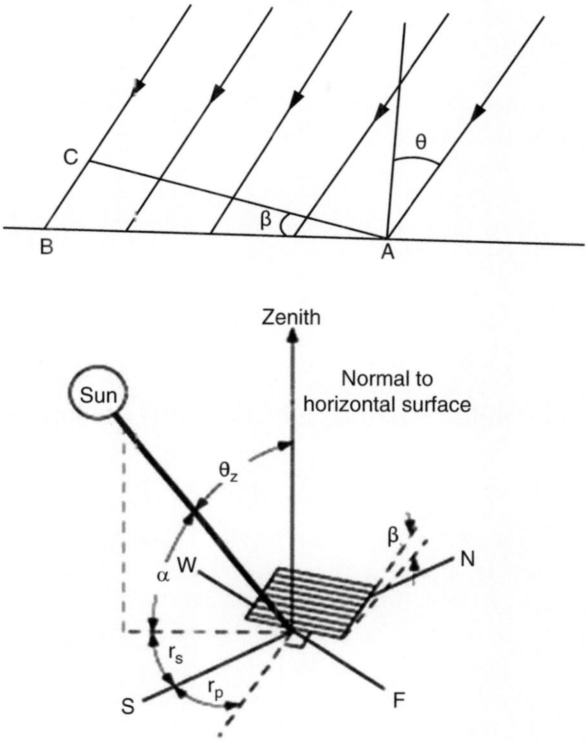

$$\beta = 90^\circ$$
$$\cos\,\theta = \cos\,\alpha\,\cos\,(r_s - r_p)$$

2. Horizontal surface

$$\beta = 0^\circ$$
$$\cos\,\theta = \sin\,\alpha$$

3. Surface facing due south

$$r_p = 0^\circ$$
$$\cos\,\theta = \cos\,\alpha\,\cos\,r_s\,\sin\,\beta + \sin\,\alpha\,\cos\,\beta$$

4. Vertical surface facing due south

$$\beta = 90^\circ, r_p = 0^\circ$$
$$\cos\,\theta = \cos\,\alpha\,\cos\,r_s$$

Some authors (Duffie and Beckman 1980; Sukhatme 1984) have used a simple formulation as an alternative to Eq. (2.2), which is as below:

$$\begin{aligned}\cos\theta = \;& \sin\,\phi \sin\delta\cos\beta - \sin\delta\cos\phi\cos r_p \sin\beta \\ & + \cos\phi\cos\delta\cos\omega\cos\beta + \cos\delta\sin\phi\sin\beta\cos r_p\cos\omega \\ & + \cos\delta\sin\beta\sin r_p\sin\omega\end{aligned}$$

2.5 Solar Radiation Measuring Instruments

(i) **Pyranometer**

Pyranometer works on the principle of thermoelectric effect. It is used to measure the global radiation and when shaded from beam radiation by using shading ring it measures the diffused radiation. A precision pyranometer is designed to respond to radiation of all wavelengths to accurately measure the total power in the incident spectrum. It contains a thermopile whose sensitive surface consists of circular, blackened, hot junctions, exposed to sun, the cold junctions being completely shaded. The temperature difference between the hot and cold junctions is the function of radiation falling on the sensitive surface. The sensing element (thermopile) is covered by two concentric hemispherical glass domes to shield it from wind and rain.

This also reduces the convection currents. A radiation shield surrounding the outer dome and coplanar with the sensing element, prevents direct solar radiation from heating the base of the instruments.

(ii) **Pyrheliometer**

It measures beam radiation from the sun at normal incidence and works on the principle of thermoelectric effect using thermopile having sensitivity 8–10 μV/W/m^2. It uses a long collimator tube to collect beam radiation whose field of view is limited to a solid angle of 5.5°. It often uses two axes tracker for continuous recordings.

(iii) **Sunshine recorder**

This instrument measures the duration in hours of bright sunshine during the course of a day. It essentially consists of a glass sphere mounted on its axis parallel to that of earth within a spherical section. The bowl and glass sphere are arranged in such a way that the sun's rays are focused sharply at a spot on a card held in a groove in the bowl. The card is prepared from a special paper bearing time scale. As the sun's elevation changes, the focused bright sunshine burns a path along this paper. The length of the trace thus obtained on the paper is the measure of the duration of the bright sunshine.

2.6 Solar Radiation on a Tilted Surface

The total radiation incident on a surface consists of

(i) Beam or direct radiation
(ii) Diffuse or scattered radiation
(iii) Solar radiation reflected from the ground or surroundings.

The solar radiation that reaches the earth's surface directly without being absorbed or scattered is called beam or direct radiation. The direct radiation produces a shadow when interrupted by an opaque object.

Diffuse radiation is that radiation received from sun after its direction has been changed by reflection and scattered by the atmospheric molecules and by dust particles. The diffuse radiation is received by earth surface from all directions.

The practical distinction between the two components is that only the beam radiation can be focused. Even on a clear day there is some diffuse radiation.

Normally the beam radiation and diffuse radiations on horizontal surface are recorded. After knowing beam and diffuse radiations on horizontal surface, Liu and Jordan (1962) have given a formula to evaluate total radiation on tilted surface.

$$I_T = I_{bh}R_b + I_{dh}R_d + \rho R_r(I_{bh} + I_{dh})$$

where R_b, R_d and R_r are known as conversion factors for beam, diffuse and reflected components respectively. I_{bh} and I_{dh} are the beam and diffuse components respectively on horizontal surface.

$$\begin{aligned} \rho &= \text{reflection coefficient of ground} \\ &= 0.2 \text{ for concrete or glass surface} \\ &= 0.7 \text{ for snowcovered surface} \end{aligned}$$

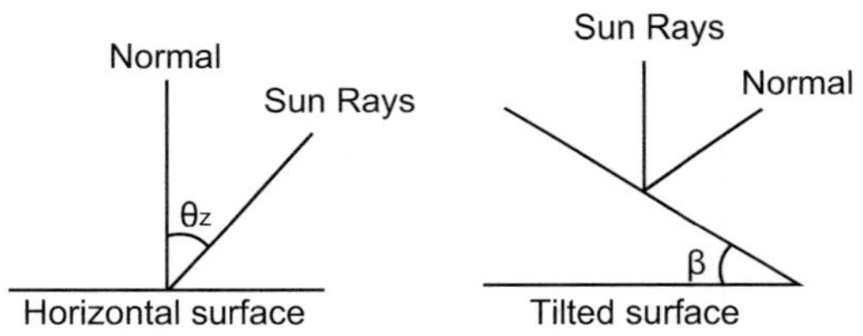

For horizontal surface

The beam radiation of horizontal surface $I_{bh} = I_N \cos \theta_z$.
where I_N is Sun rays and given by

$$I_N = \frac{I_{bh}}{\cos \theta_z}$$

For tilted surface

$$\begin{aligned} I_{bt} &= I_N \cos \theta_t \\ &= \frac{I_{bh}}{\cos \theta_z} \times \cos \theta_t \end{aligned}$$

Or,

$$I_{bt} = I_{bh} \frac{\cos \theta_t}{\cos \theta_Z}$$

But

$$\cos\,\theta_z = \cos\,\phi\,\cos\,\delta\,\cos\,\omega + \sin\,\phi\,\sin\,\delta$$
$$\cos\,\theta_t = \cos\,(\phi - \beta)\,\cos\,\delta\,\cos\,\omega + \sin\,(\phi - \beta)\,\sin\,\delta$$

where

ϕ = Latitude of the site
β = Optimum tilt angle
ω = Sun hour angle
δ = Solar declination angle
θ_t = Zenith angle on tilted surface

Thus

$$R_b = \frac{\cos\theta_t}{\cos\theta_z}$$

It is defined as the ratio of flux of beam radiation incident on tilted surface to that on a horizontal surface.

R_d: It is defined as the ratio of flux of diffuse radiation incident on tilted surface to that on a horizontal surface. It is given by

$$R_d = \frac{1 + \cos\beta}{2}$$

R_r: The reflected components come from the ground and other surrounding objects

$$R_r = \frac{1 - \cos\beta}{2}$$

Thus

$$I_{\mathrm{T}} = I_{bh}\frac{\cos\theta_t}{\cos\theta_z} + I_{dh}\frac{(1 + \cos\beta)}{2} + \rho(I_{bh} + I_{dh})\frac{(1 - \cos\beta)}{2}$$

2.7 Estimation of Average Solar Radiation

Measured solar radiation data are the best source of information for estimating average radiation on a surface relevant to solar energy utilization. In absence of such data one uses empirical relationship based on data on average hours of sunshine or average percent of possible sunshine hours; the sunshine data is widely available and is collected with Campbell Stories Instrument. Cloud cover data may also be used in this connection, but it generally embraces uncertainty resulting from visual estimates.

Solar radiation data are the basic requirement in the modelling and simulation of various energy and environment systems. For example, the global, diffuse and direct solar radiation data at horizontal and tilted surfaces is an essential element in modelling and performance evaluation of various solar systems which include:

1. Evaluation of monthly mean hourly performance during the year-round cycle on tilted flat plate collector and photovoltaic arrays.
2. The energy needs of a building wherein a large number of factors related to climate and structural configurations control the indoor temperature. The effectiveness of these factors is highly intricate to be assessed independently in a physical model.
3. Determination of solar fraction of hybrid systems and design of control logic for the operation of solar power plants.

In practice the solar radiation data is generally variable and enormously inconsistent (Sfetsos and Coonick 2000; Dorvlo et al. 2002). These data are often not available for the desired locations and are not of the required accuracy. Several empirical models have therefore been developed for the estimation of solar radiation at different geographical and meteorological conditions (Reddy 1971; Hottel 1976; Sabbagh et al. 1977; Barbaro et al. 1978; Goh 1979; El-Nashar 1981; Ogeman et al. 1984; Supit and Van Kappel 1998). These empirical relations rely on easily available meteorological parameters such as atmospheric temperature and mean duration sunshine per hour, relative humidity and total rainfall.

In recent years several authors have proposed and tested the efficacy of artificial neural network modelling for estimation of solar radiation. For example, Kaushika et al. (2014) have presented an artificial neural network model based on interrelationship of direct, diffuse and global solar radiations. In this paper, an explicit computational model based on multi output frame of neural network is developed. The approach involves the estimation of solar transmission through clear atmosphere (simple clear day model) and its modification by the allowance of weather phenomena through ANN analysis. The model is rigorous as it simultaneously considers the global, diffuse and direct components of solar radiation as target variables in the analysis. Computational algorithm is as follows:

As a first approximation, solar beam radiation on earth under clear sky condition can be estimated from the consideration of atmospheric transmittance, sun-earth geometry and climate type following the simple model due to Hottel. The solar beam radiation relation is

$$S_{bn} = I_a\left(a_n + a_1 e^{-k \sec \theta z}\right)$$

The term within the brackets is all wavelength transmittance of solar radiation through the atmosphere. It is a function of time of the day, day of the year and location on earth. I_a is the extra-terrestrial radiation and can be expressed as

$$I_a = I_{sc}\left[1 + 0.033\cos\left(\frac{360n}{365}\right)\right]$$

where I_{sc} is the solar constant 1367 W/m^2, and n is number of day ($n = 31$ for January 31) and θ_z is solar zenith angle or solar incident angle on a horizontal plane.

The beam radiation falling on a horizontal surface is therefore given by

$$S_{bh} = S_{bn}\cos\,\theta_z$$

For the estimation of diffuse component of solar radiation falling on a horizontal surface (S_{dh}), Liu and Jordan (1960) have given an expression in terms of atmospheric transmittance for beam radiation

$$S_{dh} = I_a\cos\theta_z\left[0.2710 - 0.2939\left(a_0 + a_1\,e^{-k\ \sec\ \theta_z}\right)\right]$$

This expression is rather approximate and may be used as guide in search of experimental relationship. So the global radiation may be expressed as

$$S_{gh} = S_{bh} + S_{dh}$$

The above equations may be used to estimate the values of direct, diffuse and global radiations. The deviations of these estimates from observations may be considered due to climate and weather phenomena and are characterized by the atmospheric clearness indices as follows:

$$\text{Atmospheric clearness index} = \frac{\text{Observed value of solar radiation}}{\text{Calculated value of solar radiation}}$$

A value compatible with observed values may therefore be obtained by multiplying the atmospheric clearness index with the value calculated from simple clear day model. The clearness index value for direct, diffuse and global radiation have been determined by Kaushika et al. (2014) by using Artificial Neural Network based on long-term solar radiation data in Indian region. A contour map of these determinations for typical hours of a day is shown in Fig. 2.4.

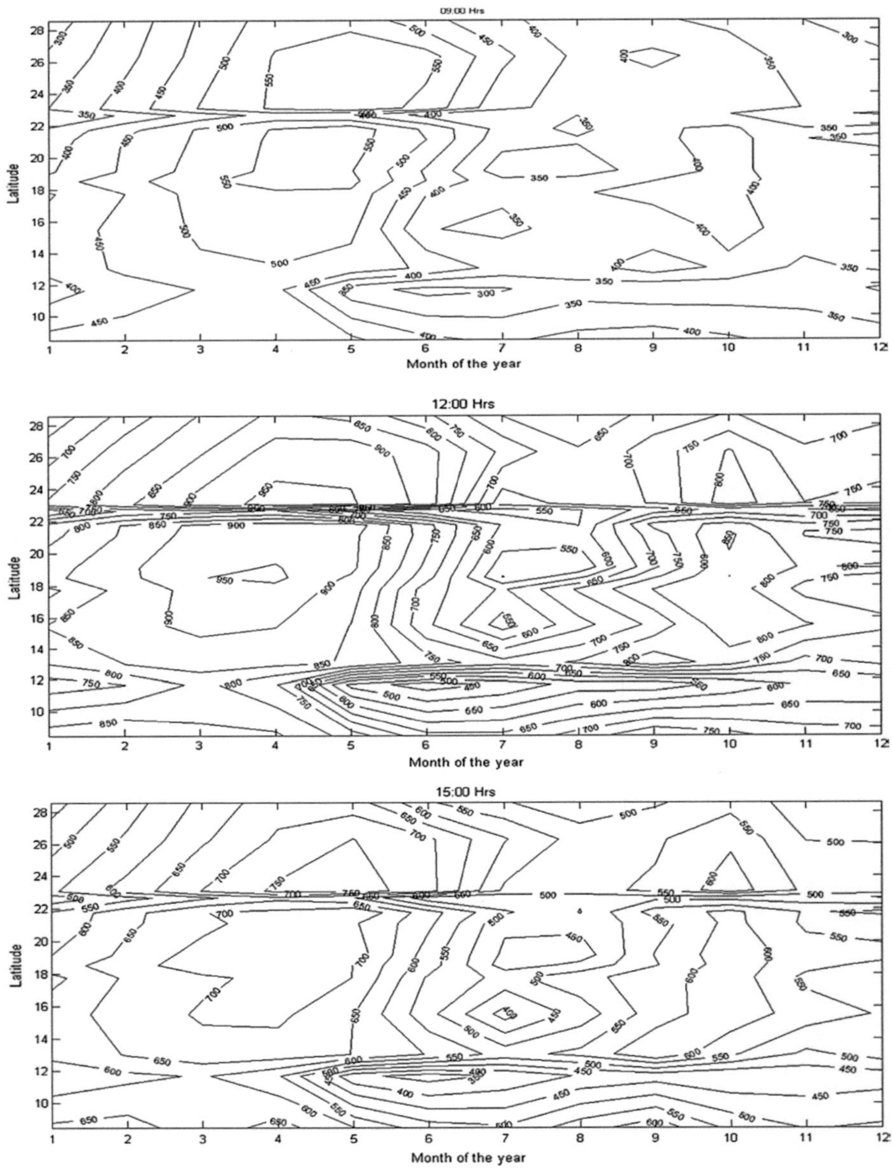

Fig. 2.4 Contour maps of clearness indices of global solar radiation for typical hours of a day for Indian region

Chapter 3
Fundamentals of Photovoltaic Generation: A Review

3.1 Introduction

Photovoltaic generation of electricity is caused by the radiation separating positive and negative charges in absorbing material. If the electric field is present these charges can generate an electromotive field and produce a current for use in an external circuit. The fundamentals of photovoltaic generation, therefore, involve the knowledge of science of materials and the photovoltaic effect therein which in turn involve such basic concepts as atomic bonding, metals, semiconductors and insulators, crystal structure (silicon), band gap energy, built-in electrostatic field, PN junction, interaction of matter and radiation. In this chapter we present a review of these concepts.

3.2 Atomic Bonding

The basis of atomic bonding is the structure of atoms which states that atoms are formed with nucleus (positive charge) in the centre and electrons (negative charge) orbiting around the nucleus at different energy levels or shells. In solids the atoms are closely packed and they create interatomic bond which is often referred to as a chemical bond. In this circumstance the atoms try to obtain a full outer shell. There is an important principle regarding electrons filling up shells: the Pauli's exclusion principle which states that two electrons cannot occupy the same energy level. The mode of filling up of shells determines the chemical, electrical and optical behaviour of the matter. In chemical bonds different or identical atoms come together for mutual minimization of electron energies and shell filling. There are four main types of bonding as follows:

N. D. Kaushika et al., *Solar Photovoltaics*,
https://doi.org/10.1007/978-3-319-72404-1_3

3.2.1 Ionic Bonding

This bonding occurs when two atoms of different species come together and the outer most electron of one atomic species is donated to other to fill its outer orbit. Consequently, the species involved in this bond are positively and negatively charged ions and they attract each other. An example of ionic bonding is NaCl (Sodium chloride).

3.2.2 Metallic Bonding

Metallic bonding occurs in metals. In this bonding outer-most electrons of atom are not accepted by any other atoms and move freely in the solid. The electron cloud around the nuclei actually acts as glue that holds the lattice of atoms together. The mobile electrons are responsible for the good conduction of electricity in metals such as silver which has conductivity about almost 6×10^7 mho/m.

3.2.3 Covalent Bonding

In this bonding none of the outer-most electrons are able to move freely but are shared amongst atoms involved in the bond. Diamonds are an example of carbon atoms in a covalent bond. This is responsible for the electrical insulation character of diamonds and fused quartz (thermal conductivity less than 2×10^{-17} mho/m). In case of silicon and germanium, the covalent bonding is rather weak and some electrons may be released to move freely when sufficient thermal energy is provided. It gives the covalent bond atoms the character of semiconductor. Silicon and germanium have crystalline structure. Their atoms are arranged in an ordered array known as crystal lattice. These materials have four valence electrons in their outer shell. The neighbouring atoms form covalent bonds by sharing four electrons with each other so as to achieve inert gas structure.

In the case of intrinsic semiconductor, the covalent bonds have to be broken thermally to provide electrons for conduction.

3.2.4 Van der Waals Bonding

It occurs in atomic species where the atoms behave like an instantaneous dipole and it inducts a very small but opposite dipole moment on another atom. Consequently a very weak bond of attraction is created. It is found in noble gases such as argon, crypton etc.

3.3 Crystal Structure

Materials can be broadly classified into two categories: crystalline and non-crystalline solids.

A solid in general is said to be a crystal if the constituent particles (atoms, ions or molecules) are arranged in a three dimensional periodic manner. In a crystalline material the arrangement of atoms is in periodically repeating pattern whereas no such regularity of arrangement is found in a non-crystalline material. A crystalline solid can be either a single crystal, where the entire solid consists of only one crystal. It has well-defined crystal structure, higher density due to closed packing of atoms and exhibits sharp diffraction pattern upon irradiation.

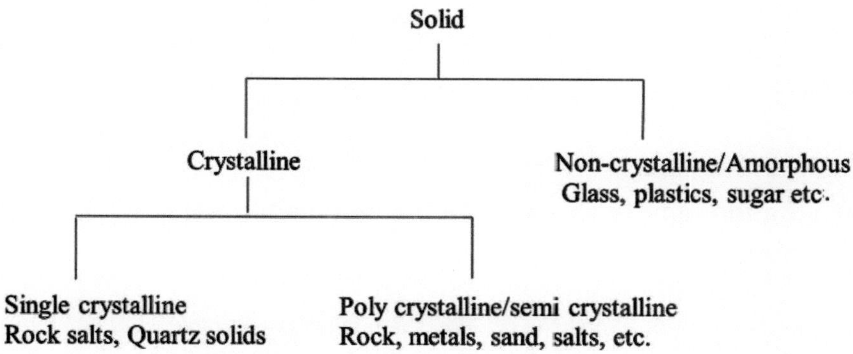

Non-crystalline solid often possesses entangled chain without periodicity. It has lower density as the packing of atoms is zigzag, does not show sharp diffraction pattern and melts over a range of temperatures. However in both forms, the coordination number is almost the same. Formation of non-crystalline structure is characterized by several factors as listed below:

(i) Formation of one-dimensional chain molecule.
(ii) Formation of two-dimensional sheet molecules.
(iii) Non-formation of three-dimensional primary bond.
(iv) Absence of primary bonds in all the directions.
(v) Non-parallel, entangled chain configuration.
(vi) Weak secondary bond.
(vii) Open network of the atomic packing.

Typical crystalline and amorphous structures are illustrated in Fig. 3.1.

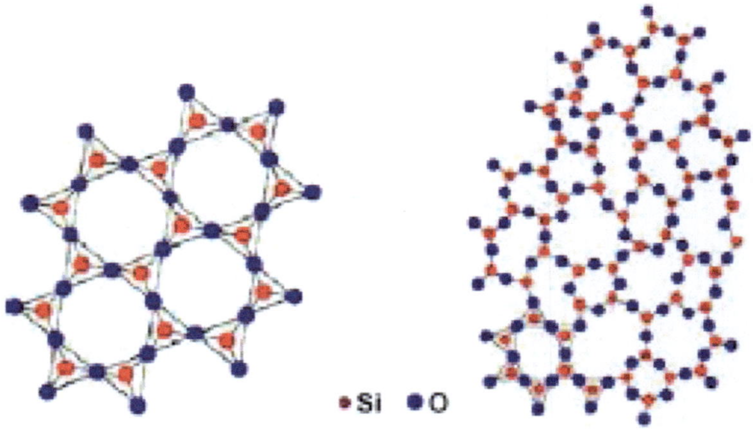

Fig. 3.1 Crystalline and amorphous structures of solid

3.4 Band Gap Structure of Metal, Semiconductors and Insulators

The characteristics of a solid can be understood in terms of band energy diagram as shown in Fig. 3.2. Its energy band structure and the manner in which electrons occupy the energy-band are central to the electrical properties of a solid. The energy levels split into *N* levels of energy where N is the number of atoms in the crystals. The electrons in the outer-most orbit of an atom are known as valence electrons. Under normal condition of the atom, valence band corresponds to the electrons of highest energy. This band may be filled completely or partially but never empty. The energy band which corresponds to the valence electrons is called valence band.

The next higher permitted energy band is called the conduction band. It refers to allowed energy of bound electrons. It may be either empty or partially filled with electrons. The gap between the valence band and conduction band is called forbidden energy gap. It is denoted by *Eg* and may be defined as the energy difference between the most energetic bound electrons and minimum allowed energy of conduction electrons governed by quantum mechanical effect. So forbidden energy gap is a region in which no electron can stay as there is no allowed energy state. The width of the forbidden energy gap represents the bondage of valence electrons to the atom. The greater the forbidden energy gap more tightly the valence electrons are

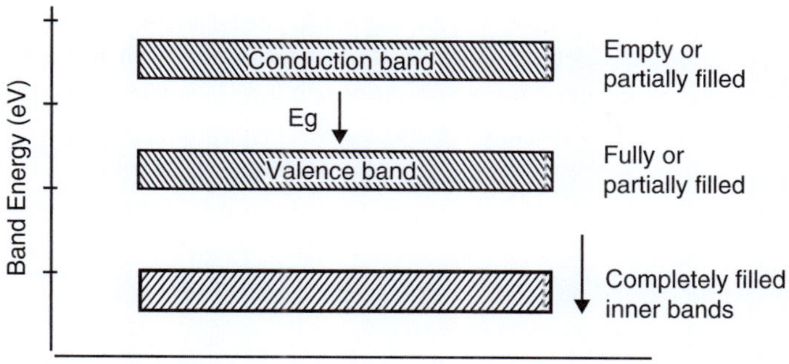

Fig. 3.2 Energy band diagram of a solid

Fig. 3.3 Energy-band diagram for three types of material: Metals, Semiconductors and Insulators

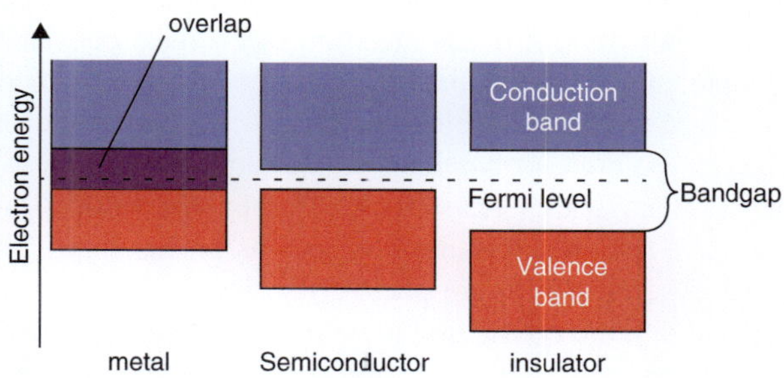

bound to the nucleus. When valence electron absorbs enough energy, it jumps across the forbidden energy gap and enters the conduction band.

Fermi Level refers to the highest occupied molecular energy orbital at absolute zero. It is usually found at the centre between the valence and conduction bands. The particles in this state each have their own quantum states and generally do not interact with each other. When the temperature begins to rise above absolute zero, these particles will begin to occupy states above the Fermi level and states below the Fermi level become unoccupied.

Depending upon the band theory, all solids can be classified as metals, semiconductors and insulators (Fig. 3.3).

In metals (conductors) there is no forbidden energy gap between the valence bands and conduction bands. They overlap each other. The valence–band energy is same as conduction band energy. It is very easy for a valence electron to become a conduction electron. Therefore, without supplying any additional energy, a metal already contains a large number of free electrons and works as a good conductor.

Insulator has a very wide forbidden-energy gap; because of this, it is impossible for an electron in the valence band to enter into the conduction band. At room temperature, an insulator does not conduct because there are no conduction electrons in it. However, it conducts only if its temperature is very high or a very high voltage is applied across it.

Semiconductor material is one whose electrical properties lie in between those of insulators and good conductors. The forbidden energy gap is of the order of 1 eV. The energy provided by the heat at room temperature is sufficient to lift electrons from the valence band to conduction band.

3.4.1 Commonly Used Semiconductor Materials

A large number of tetravalent (group IV of Periodic Table) materials such as carbon in diamond state, silicon, germanium and gray tin are semiconductors. The formation of allowed energy bands is related to the overlapping of atomic orbitals in the crystal. The concept of band gap is important in understanding the behaviour of these semiconductors. It needs quantum mechanical considerations. However, there is a simple intuitive explanation based on the fact that all of valence electrons are involved in strong covalent bonds that require some energy to break. The minimum energy required for breaking the covalent bond in these materials is 7 eV, 1.12 eV, 0.75 eV and 0.1 eV respectively. Carbon in diamond state, having forbidden energy gap of 7 eV, behaves more or less as an insulator. Silicon and germanium need 1.12 eV and 0.72 eV respectively to break a covalent bond; therefore, silicon and germanium are considered to be most suitable semiconductor materials.

Silicon is most common and widespread element on the earth. The word silicon has origin in the Latin word silex which means flint. It does not occur free in nature. It is always found as a compound and almost always as an oxide. The simplest compound of silicon and oxygen is silicon dioxide which is commonly known as silica or sand. Silicon is obtained from silicon dioxide. Silicon is separated from oxygen by the use of a reducing agent like coke which is one of the cheapest reducing agent. Silicon dioxide is heated with coke in an electric furnace

$$SiO_2 + 2C \rightarrow Si + 2CO \uparrow$$

Silicon also occurs in nature as quartz, flint and precious stones like opal, agale etc.

Silicon belongs to group IV of the Periodic Table and placed below carbon in the table. The atomic number and atomic mass of silicon are 14 and 28 respectively and having 14 protons, 14 neutrons and 14 electrons. The electronic configuration of silicon is 2, 8 and 4. Like carbon it has four electrons in its outermost shell or valence shell. Hence it is tetravalent.

Properties of Silicon

- Silicon is a grey, hard and shining substance with density 2.33 gm/cm^3 at room temperature. Band gap energy 1.12 eV at 300 K.
- It melts at 1410 °C and boils at 365 °C.

- It is a crystalline solid and its atomic structure is similar to that of diamond. Each silicon atom is surrounded by four other silicon atoms. All the silicon atoms are held together by a network of covalent bonds.

3.4.2 Types of Semiconductors: Doping

The classification of semiconductors can be done as shown in Fig. 3.4.

An intrinsic semiconductor is one which is made of the semiconductor material in its extremely pure form. Intrinsic semiconductor materials are: pure germanium and silicon which have forbidden energy gap of 0.72 eV and 1.1 eV respectively. The energy gap is so small that even at ordinary room temperature, electrons jump across the small energy gap between the valence and conduction bands creating positively-charged holes in the valence band i.e. an intrinsic semiconductor has equal number of conduction electrons and holes. Although the holes exist in the valence band, they are similar to conduction electrons as they serve as mobile charge carriers. The method of movement is little different. Essentially a hole moves through the crystal when electrons from neighbouring silicon atom move into fill the hole and replace the missing electron. When this happens, the neighbouring silicon atom then inherits the broken bond and the process advances. Both holes and electrons move with quite similar dynamics although they have opposite charges and exist in different electronic bands.

In an intrinsic semiconductor, Fermi level lies exactly in the middle of the forbidden energy gap i.e. midway between the conduction and valence bands (Fig. 3.5).

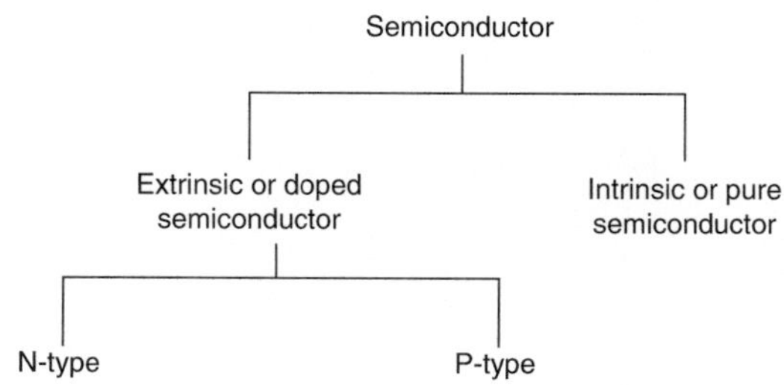

Fig. 3.4 Types of semiconductor materials

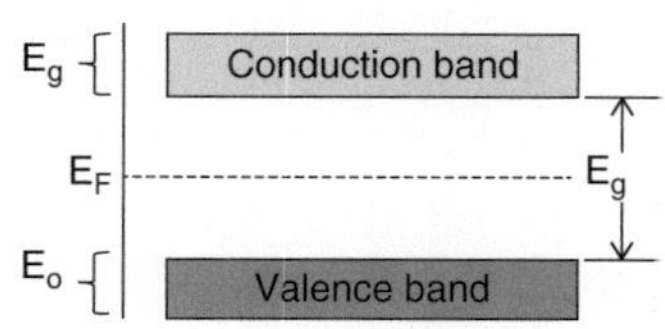

Fig. 3.5 Fermi level in intrinsic semiconductor

The Fermi level may be defined as the energy which corresponds to the centre of gravity of conduction electrons and holes weighed according to their densities. The number of electrons in conduction band is n_c and given by

$$n_c = NP(E_g)$$

where N is total number of electrons in both conduction and valence bands and $P\,(E_g)$ represents the probability of an electron having energy E_g.

The value of $P(E_g)$ can be found from Fermi Dirac probability distribution function and is given by (e.g. Theraja and Theraja 1988)

$$P(E) = \frac{1}{\exp\left\{\frac{E-E_F}{kT}\right\} + 1}$$

where k is Boltzmann's constant, T is the absolute temperature, $P(E)$ is the probability of finding an electron having any particular value of energy E and E_F is Fermi level.

$$\therefore P(E_g) = \frac{1}{\exp\left\{\frac{(E_g-E_F)}{kT}\right\} + 1} \therefore n_c = \frac{N}{\exp\left\{\frac{(E_g-E_F)}{kT}\right\} + 1}$$

The number of electrons in valence band n_v is given by

$$N_v = NP(0)$$

where N is total number of electrons in both conduction and valence bands and $P\,(0)$ represents the probability of an electron having zero energy in valence band.

The value of $P(0)$ can be found by putting E equal to zero in Fermi Dirac probability distribution function and is

$$P(0) = \frac{1}{\exp\left\{\frac{0-E_F}{kT}\right\} + 1} = \frac{1}{\exp\left\{\frac{-E_F}{kT}\right\} + 1}. \text{ Hence } n_v = \frac{N}{\exp\left\{\frac{-E_F}{k\,T}\right\} + 1}$$

The total number of electrons $N = n_c + n_v$.

$$\text{Or, } N = \frac{N}{\exp\left\{\frac{(E_g-E_F)}{kT}\right\} + 1} + \frac{N}{\exp\left\{\frac{-E_F}{kT}\right\} + 1}$$

which on simplification gives $E_F = \frac{E_g}{2}$.

Extrinsic or doped semiconductors are those intrinsic semiconductors to which some suitable impurity or doping agent dopant has been added in extremely small amount—about 1 part in 10^8. The usual doping agents are pentavalent atoms having five valence electrons (antimony, arsenic, and phosphorus) or trivalent atoms having three valence electrons (indium, gallium, aluminium, boron). Depending upon type of doping material used, extrinsic semiconductors can be divided into two cases:

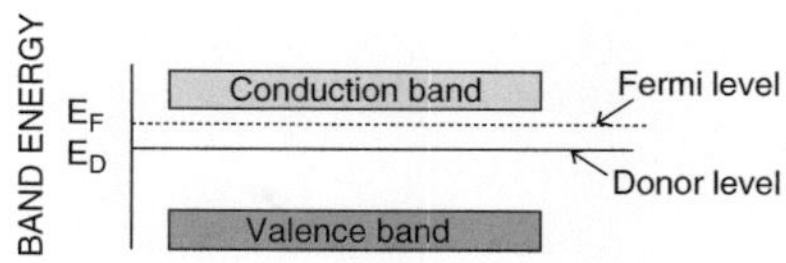

Fig. 3.6 Fermi level in *N*-type extrinsic semiconductor

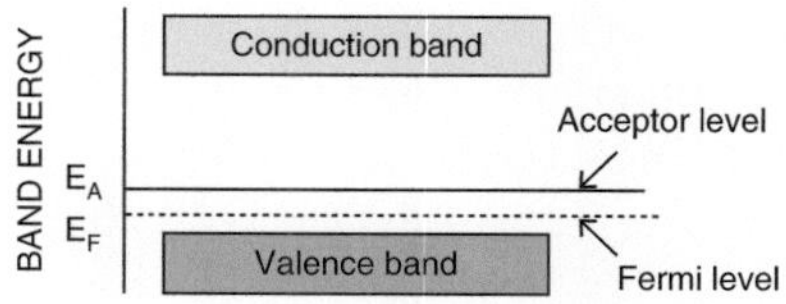

Fig. 3.7 Fermi level in *P*-type extrinsic semiconductor

- *N*-type extrinsic semiconductor
- *P*-type extrinsic semiconductor

In *N*-type extrinsic semiconductor, a pentavalent material like antimony is added to pure silicon crystal. Each antimony atom forms covalent bonds with the surrounding four silicon atoms. The fifth electron can be easily excited from the valence bond to the conduction band by the application of electric field or increase in thermal energy. Thus, antimony is called donor impurity and makes the pure silicon an *N*-type extrinsic semiconductor. The addition of antimony increases the number of conduction electrons. Hence, Fermi level shifts towards the conduction band because of increased concentration of electrons in the conduction band (Fig. 3.6).

In *N*-type semiconductors electrons are the majority charge carriers while holes are the minority charge carriers.

In *P*-type extrinsic semiconductor, a trivalent material like boron is added to a pure silicon crystal. The three valence electrons of boron atom form covalent bonds with four surrounding silicon atoms. The fourth silicon atom is unable to form a covalent bond with boron atom because boron atom does not have the fourth electron in its valence orbit and gives rise to a hole. Thus boron causes as many positive holes in a silicon crystal as there are boron atoms thereby producing a *P*-type extrinsic semiconductor. In this type of semiconductor, conduction is by the movement of holes in the valence band; hence holes are the majority charge carriers whereas electrons are minority charge carriers. Since concentration of holes in the valence band is more than the concentration of electrons in the conduction band, Fermi level shifts nearer to the valence band and acceptor level lies immediately above the Fermi level (Fig. 3.7).

Conduction is by means of hole movement at the top of valence band, the acceptor level readily accepting electrons from the valence band.

3.4.3 Drift, Diffusion and Recombination Currents

Movement of charge carriers occur in semiconductor materials through following mechanisms:

- Drifting of charge carriers under the influence of applied electric field and
- Diffusion of charge carriers from a high charge density region to low charge density region.

Above 0°K when no electric field to the semiconductor crystal is applied, the motion of the charge carriers is random and they collide with each other and fixed ions. Because of this the net average velocity of these charge carriers in any given direction is zero and hence no current exists in the crystal.

When electric field is applied to the semiconductor crystal, the motion of charge carriers is directed. Due to this the net average velocity, known as *drift* velocity, is in the direction of applied electric field. Since the holes and electrons move in opposite direction because of their opposite charges but produce a current in the same direction. The drift velocity is proportional to electric field strength and is given as

$$v = \mu E$$

where v = drift velocity, E = applied electric field strength and μ = proportionality constant, known as mobility.

The current density due to electron drift is $e\,\mu_e\,n\,E$ where μ_e is electron mobility and n electron density.

The current density due to hole drift is $e\,\mu_h\,p\,E$ where p is hole density. The total current density due to electron and hole drift is

$$J = e\mu_e n\,E + e\mu_h p\,E = e(n\mu_e + p\mu_h)E$$

where μ_e = electron mobility, n = current density, μ_h = hole mobility, h = hole density and J = total current density.

Diffusion is gradual flow of charge carriers from a region of high density to a region of low density due to non-uniform distribution of charge carriers in a semiconductor crystal. It is a force-free process and leads to an electric current without the benefit of an applied electric field. The diffusion of charge carriers is proportional to the carrier density gradient, and the constant of proportionality is diffusion constant or diffusion coefficient. The current density due to electron diffusion is

$$J_e = e\ D_e\frac{dn}{dx}\ \text{and due to hole diffusion}$$

$$J_h = -e\ D_h\frac{dp}{dx}$$

where D_e, D_h = electron and hole diffusion constants respectively ofunit, $\frac{\text{m}^2}{\text{s}}$, $\frac{dn}{dx}$ = electron density gradient and $\frac{dp}{dx}$ hole density gradient.

In semiconductor materials, drift and diffusion processes occur simultaneously and, therefore, the expression for total electron and hole current densities become

$$J_e = e\mu_e nE + eD_e \frac{dn}{dx} \frac{\text{Amp}}{\text{m}^2}$$
$$and\, J_h = e\mu_h pE - eD_h \frac{dp}{dx} \frac{\text{Amp}}{\text{m}^2}$$

Apart from drift and diffusion, a third phenomenon called *recombination* occurs in semiconductor crystals. The recombination is a result of collision of an electron with a hole and the recombination rate is directly proportional to the carrier concentration. It also depends on the carrier life time that is defined as time for which, on an average, a charge carrier will exist before recombination. Its value varies from nanoseconds to microseconds. The total carrier flow in a semiconductor is sum of the three flows as discussed above. The current in the semiconductor is the sum of the electrons and hole currents.

3.5 P-N Junction

When *n*-type material (characterized by electrons as majority carriers and holes as minority carriers) is in metallurgical contact with a piece of *p*-type material (characterized by holes as majority carriers and electrons as minority carriers) a *p-n* junction is formed (Fig. 3.8). This type of junction is commonly encountered in many semiconductor devices such as transistors, diodes and photodiodes. The solar cell is a large area photodiode and understanding of this junction is important to understand the operation of solar cell.

The density of donor and acceptor atoms is equal in the junction plane. The conduction electrons in *N*-type region and holes in *P*-type region are highly mobile and they possess thermal energy and exhibit kinetic motion that causes them to drift randomly through the crystal. Consequently, at the junction, conduction electrons from donor atoms in the *n*-region diffuse into the *p*-region and combine with acceptor atoms, producing a negatively charged impurity atoms. The opposite action also takes place: conduction holes from acceptor atoms in the *p*-region diffuse into the *n*-region and combining with electrons, producing a positively charged impurity atoms. The net result of this movement is the disappearance of conduction electrons and holes from the vicinity of the junction and establishment of reverse electric field which is positive on *N*-type side and negative on *P*-type side. The area in which this field is set up is called depletion layer (Fig. 3.9).

If a majority carrier, either an electron or a hole, tries to cross into depletion layer either it can be trapped by the row of fixed ions of opposite sign or it may succeed in entering the depletion layer where it will be repelled by the row of similar-charged impurity ions guarding the other region. At the end, an equilibrium condition is reached when depletion layer has widened to such an extent that no electrons or holes can cross the *P-N* junction. Because of this charge separation, an electric

Fig. 3.8 *P-N* junction

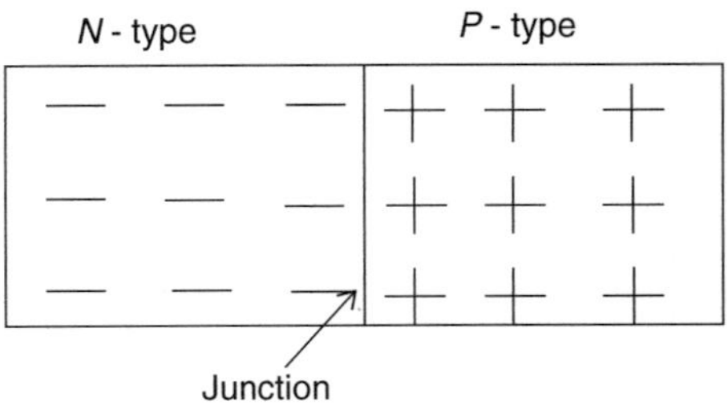

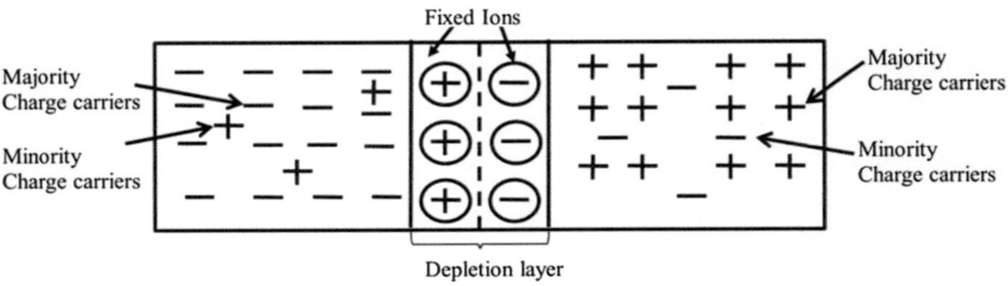

Fig. 3.9 Depletion layer in *p-n* junction

potential difference is established across the junction known as *junction* or *barrier* potential (Fig. 3.10) which depends on doping density, electronic charge and temperature. This junction potential, at room temperature, is about 0.3 V and 0.7 V for germanium and silicon respectively (e.g. Theraja and Theraja 1988). For any junction, the doping density and electronic charge are constant, thus making barrier potential dependent on temperature. As temperature increases, more minority charge carriers are generated leading to an increase in drifting across the junction. It has been found that the barrier potential is decreased by about 2 mV per degree centigrade for germanium and silicon.

Figure 3.11 shows the energy bands of the trivalent impurity atoms in the *P*-region at a slightly higher level than those of the pentavalent impurity atoms in *N*-region at the instant of junction formation. This is because of higher core attraction for valence electrons in pentavalent atoms. Therefore, trivalent valence electrons are at a higher energy level. However, there is some overlap between respective bands of the two regions. Due to this reason, some high energy electrons near the top of *N*-region conduction band diffuse into the lower part of the *P*-region conduction band. As diffusion continues, depletion layer begins to form and energy bands in *N*-region shift downward due to loss of high energy electrons.

When the top of the conduction band in *N*-region is at the same level of bottom of conduction band in *P*-region an equilibrium condition is reached with establishment of barrier potential. For an electron to diffuse across the junction, it must receive energy from the external source.

When the *P-N* function is forward biased, junction offers a low resistance and permits easy flow of electrons across the junction. In the case of reverse biased, *P-N* junction offers high resistance and no current flows across the junction.

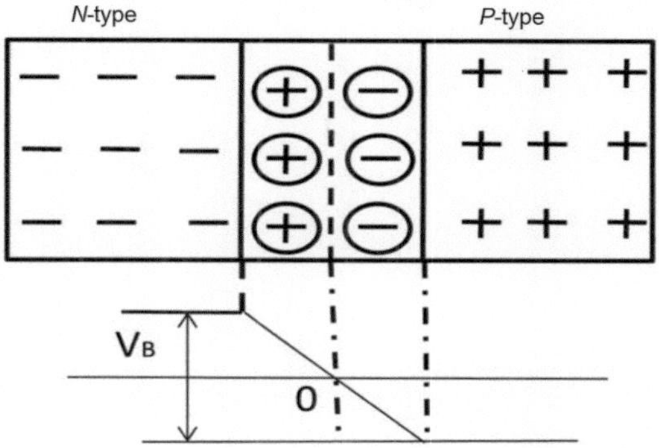

Fig. 3.10 Junction potential in *p-n* junction

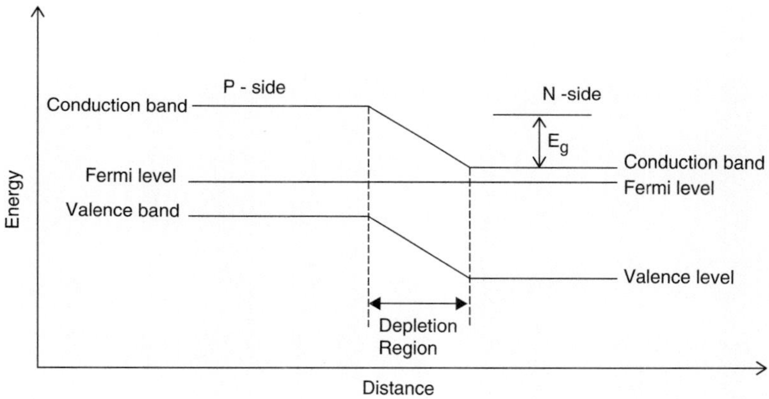

Fig. 3.11 Energy band diagram in *P-N* junction

3.6 P-N Junction Solar Cells

A solar cell is a large area junction device made from a thin (250-400 μm) wafer cut from a pure silicon crystalline block. In the cell fabrication process a *p*-type doped silicon wafer 300 μm thick is taken and the phosphorous (having five valence electrons) is diffused into the top region of slice such that about 0.5 μm of silicon wafer is converted into *n*-type which leads to the formation of *p-n* junction (Fig. 3.12).

A current collected grid is connected to the top and bottom ends of the solar cell to enable the external characteristics. The front metal contact is in the form of a narrow-fingered grid, while the back contact usually covering the entire back surface. The front contact surface has an antireflective coating to keep the photons going into the crystal instead of bouncing back. Titanium oxide and tantalum pentoxide are the materials being used as an antireflective coating. It can be square, square with cropped corners, circular or semicircular. Nowadays it is usually 100 mm square or 100 mm diameter (Fig. 3.13).

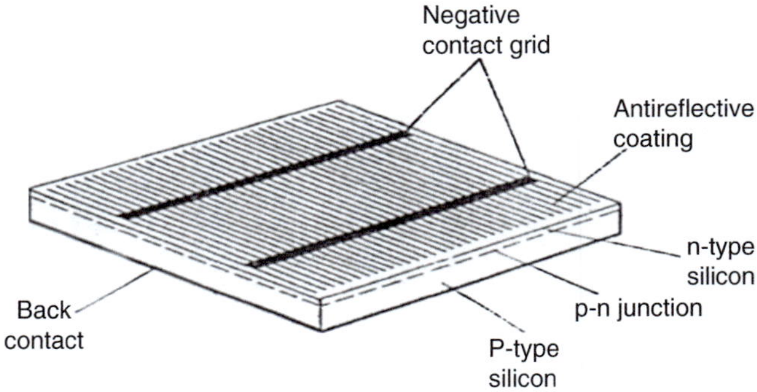

Fig. 3.12 *P-N* junction solar cell

Fig. 3.13 100 mm circular and square solar cells

When light falls on front surface, the photon energy in the excess of band gap interact with valence electron and lift them to conduction band. This movement leaves behind holes; so each photon is said to generate an electron-holes pair.

In crystalline silicon, electron-hole generation takes place throughout the thickness of the cell in concentration depending upon the irradiance and spectral composition of the light. The electrons and holes diffuse through the crystal in an effort to make an even distribution. Some recombine after a lifetime of the order of 1 ms neutralizing their charges and giving up energy in the form of heat. Others reach the junction before their life time has expired where they are separated by the reverse field, the electrons being accelerated towards the negative contact and holes towards positive contact. If the cell is connected to a load (Fig. 3.14), electrons will be pushed from the negative contact through the load to the positive contact, where they will recombine with holes. This constitutes an electric current.

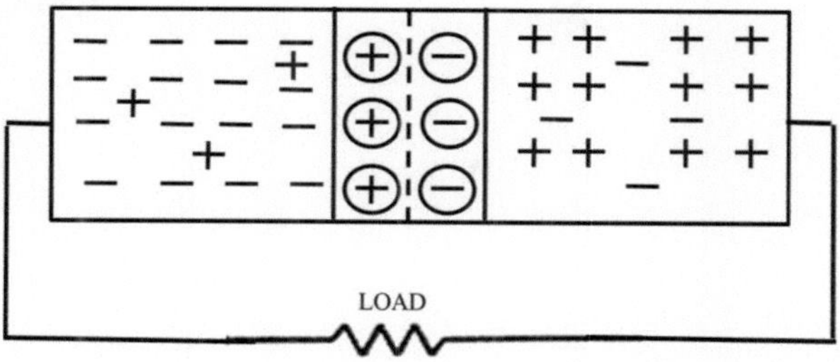

Fig. 3.14. *p-n* junction solar cell connected to a load

Chapter 4
Wafer-Based Solar Cells: Materials and Fabrication Technologies

4.1 Photoionic Processes in Solar Cells

Solar cells convert solar energy into electrical energy by photoionic process referred to as photoelectric effect. The solar radiation may be considered as discrete bundles of energy called photons. The energy (E) of photons associated with radiation of frequency (ν) and wave length (λ) may be expressed as $E = h\nu = h\ c \backslash \lambda$ where h is Planck's constant and c is the velocity of light. The radiation intensity is proportional to photon number density. As discussed in Chapter 2 the solar spectral irradiance ranges between the wave length region of 0.2–2.0 μm and has maximum invisible light region at 0.48 μm.

With the solar radiation, photoelectric effect is possible in a number of materials but the materials that have shown best performance in sunlight are the semiconductors. In a direct absorber semiconductor material, the incoming photons and electrons interact directly. The photons of energy (E_p) greater than the forbidden energy gap (E_g) are completely absorbed and the electrons are transferred to conduction band. The excess energy ($E_p - E_g$) is given up to lattice as thermal energy. For photons with energies less than E_g, the material will appear transparent (Backus 1977). Such absorption is not readily attained in indirect absorber semiconductors wherein the photon interacts with the electron as well as phonons from the lattice to provide momentum change. Silicon and germanium are indirect absorbers. Doped crystals of these materials are, therefore, used. They are of two types: (a) n type which is doped with an impurity such as phosphorous of 5 valence electrons and (b) p type, which is doped with an impurity such as boron of 3 valence electrons. After the photons are absorbed and electron hole pairs are created, the charges are to be separated. In solar cells the necessary electric field is provided by a junction formed by alloying or diffusion processes. The junction can be made inside the same semiconductor and is then referred to as homojunction in contrast to a heterojunction made between two different semiconductors such as CdS and Cu_2S. The junction can also be made between a metal and semiconductor, e.g. Si and Ag. Yet another approach due to Gover and Stella (1974) suggest a new geometry of multi junctions for solar cells.

N. D. Kaushika et al., *Solar Photovoltaics*,
https://doi.org/10.1007/978-3-319-72404-1_4

In the junction region free electrons of *n* type semiconductor neutralize holes of *p* type region. Consequently *n* region interface is left with excess positive charges and *p* region with negative charges. These charges give rise to an electric field (depletion layer) at the *p-n* interface. If the photon-produced minority carriers (holes in *n* type and electron in *p* type) are able to diffuse to depletion layer region before they recombine, they are swept across the junction and separated giving rise to a photo voltage.

4.2 Solar Cell Materials

In the pioneering work at Bell Telephone laboratory on solar cells, Chaplin et al. (1954) used crystalline silicon to fabricate a diffused junction solar cell and Reynolds et al. (1954) made a similar breakthrough with a cadmium sulphide device. Following this, several semiconductors have been tried for the fabrication of solar cells. These include silicon (crystalline, polycrystalliane, amorphous), germanium, gallium phosphide, gallium arsenide, gallium antimonide, indium phosphide, indium arsenide, indium antimonide, zinc oxide, zinc sulphide, cadmium sulphide, cadmium selenide, lead sulphide, lead selenide and cadmium telluride. The PV materials of commercial potential are given in Table 4.1 (Kazmerski and Emery 1992; Kaushika 1996). The efforts in cell fabrication technology are summarized in Fig. 4.1.

These semiconductor materials have intermediate conductivities (1–10 mho/m) between insulators (e.g. fused quartz has conductivity less than 2×10^{-17} mho/m) and good conductors (e.g. silver has conductivity about 2×10^{7} mho/m). This property results from the covalent bonding structure of their crystal lattice. Silicon has a chemical valence of 4 which implies that out of its 14 electrons, four are available to interact with other atoms. The regular structure of silicon crystal is such that each atom is firmly held in the crystal lattice by sharing its valence electrons with each of surrounding four neighbours.

Table 4.1 The PV materials with their efficiencies

Material	Typical cell voltage at open circuit (volts)	Typical cell current at short circuit (mA/cm^2)	Efficiency (%)
Mono-crystalline	0.6	35	14–15
Poly crystalline	0.6	35	12–13
Gallium arsenide	1.0	27	24
Tandem amorphous silicon (Two cells)	1.8	10	25
Amorphous silicon	0.9	15	9–10
Copper indium diselenide (CIS)	0.4	35	10
Cadmium sulphide, Cadmium telluride	0.7	25	10

For a III-IV semiconductor (e.g. gallium arsenide GaAs) the electron pair is provided by atoms of 1:1 mixture of III and IV column elements which behaves like an average lattice atom that contributes four valence electrons. The same argument applies to II and VI semiconductor e.g. cadmium sulphide. This situation is similar to that of an insulator and is in contradiction to that of metals where many electrons are free to move about to constitute an electric current upon the application of electromotive force. In terms of energy band concept or the band theory of materials, the energy of electrons can be represented by a one dimensional energy diagram consisting of allowed and forbidden bands. Conduction bands are empty in insulators and have a sea of electrons in good conductors. In semiconductors the forbidden gap (the gap between conduction and valence bands) is much narrower. Energy gaps of some common semiconductors are shown in Table 4.2.

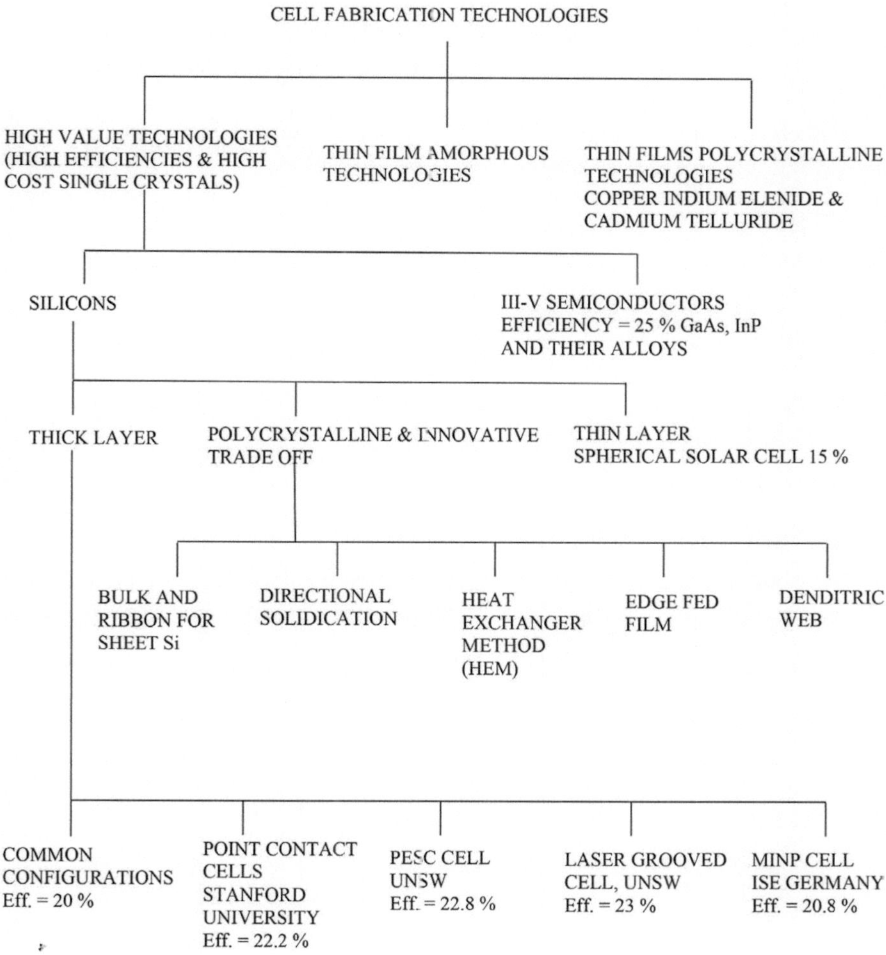

Fig. 4.1 Cell fabrication technologies (Kazmerski and Emery 1992; Kaushik 1996)

Table 4.2 Energy gaps of some semiconductors

Material	Energy gap (eV)
Crystalline silicon	1.12
Amorphous silicon	1.65
Germanium	0.66
Gallium phosphide	2.26
Gallium arsenide	1.42
Indium phosphide	0.92/1.35
Indium arsenide	0.36
Cadmium sulphide	2.6
Cadmium telluride	1.4

Some electrons may receive enough energy from sunlight or thermal agitations to jump across the influence of an electric field. The material in which the conduction is due to electrons excited from valence band to conduction band is called intrinsic semiconductor. In this case only a limited portion of the solar energy can be used to excite electrons, so there is a theoretical limit on the efficiency of photovoltaic conversion. The efficiency of photovoltaic conversion is illustrated as follows:

The energy flux E_λ represents energy per unit area per unit time per unit wavelength interval at wavelength λ. So if ν is frequency of a spectral component of radiation and h is plank's constant ($6.6260755 \times 10^{-34}$ Ws^2), the energy of each photon is $h\nu$ and the number density of photon (N_ν) in terms of solar spectral density E_ν of electromagnetic radiation is given by

$$N_\nu = \frac{E_\nu}{h\nu}$$

So, the number of photons in the spectral range ($d\nu$) is represented by

$$dN = N_\nu d\nu$$

$$\text{or, } dN = \frac{E_\nu d\nu}{h\nu}$$

The solar ionization process in solar cell requires minimum photon energy ($E_g = h\nu_g$) where E_g is band gap energy of semiconductor material. The solar spectral part of $\int_0^{\nu_g} E_\nu d\nu$or $\int_{\lambda_g}^{\infty} E_\lambda \; d\lambda$ is not being used in photon ionization and gets converted into heat.

Only the fraction of solar energy input used for carrier generation is in the frequency range greater than ν_g. So the number of photons participating in carrier generation is N_p, which is the integral of dN from ν_g to ∞ i.e.

$$N_P = \int_{\nu_g}^{\infty} \frac{E_\nu \, d\nu}{h\nu}$$

The fraction of solar energy input (E_p) actually used for charge carrier generation is $E_g N_p$ corresponding to number of photons with energy $>E_g$. So we have

$$E_g N_p = E_g \int_{\nu_g}^{\infty} \frac{E_\nu \, d\nu}{h\nu}$$

The fraction of solar energy used for photon ionization is, therefore, given by

$$\eta_c = \frac{E_g \int_{\nu_g}^{\infty} \frac{E_\nu \, d\nu}{h\nu}}{\int_0^{\infty} E_\nu \, d\nu}$$

This efficiency depends on the band gap of semiconductor material and the temperature. Silicon has an energy gap of 1.08 eV; hence, sunlight with wavelength shorter than 1.15 m can only excite electrons to conduction band. This portion of sunlight contains about 78% of total energy. Of this, only a fraction equal to 1.08 eV divided by the energy of incident photons is used to overcome the gap energy while the rest is dissipated as heat. This yields about 42% as maximum efficiency of silicon solar cells (McDaniels 1979). Based on the above consideration, a calculated curve of maximum efficiency as a function of energy gap of semiconductors is shown in Fig. 4.2.

4.3 Silicon Solar Cell Wafer Technology

The considerations of input solar spectrum, operating temperature and availability of semiconductor grade material and its processing like cutting, polishing, alloying or diffusion and contact making etc. have often favoured silicon for making solar cells. Consequently, silicon solar cells have reached the state of commercial realization. The technology of silicon solar cell wafer fabrication comprises the following steps (Green 1979):

(i) Sand to metallurgical grade silicon,
(ii) Metallurgical to semiconductor grade silicon,
(iii) Semiconductor grade silicon to single crystal silicon wafers,
(iv) Silicon wafers to solar cells, and
(v) Electrical contacts and encapsulation.

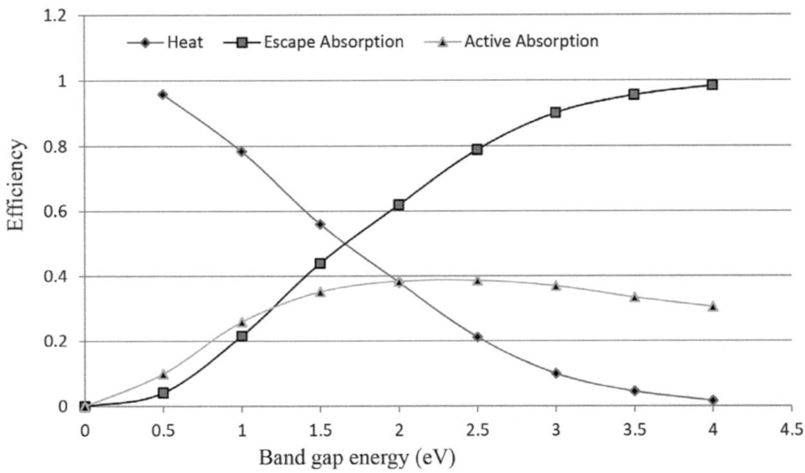

Fig. 4.2 Efficiency vs. band gap energy

4.3.1 Sand to Metallurgical Grade Silicon

In the commercial process, quartzite, an ore of silicon dioxide obtained from earth's crust is reduced in large arc furnaces by carbon (in the form of wood chips, coke and coal) resulting in silicon.

$$SiO_2 + 2C \rightarrow Si + 2CO \quad (4.1)$$

Silicon is poured into shallow troughs from the furnace and blown with oxygen or oxygen/chlorine mixture for further purification.

Its purity is around 98–99%, iron and aluminum being the major impurities.

4.3.2 Metallurgical to Semiconductor Grade Silicon

Metallurgical grade silicon is converted into a gaseous compound which is condensed and refined by fractional distillation. Ultra pure silicon is extracted from the refined product.

A bed of fine metallurgical grade silicon particles is fluidized with HCl in the presence of a Cu catalyst to promote the reaction.

$$Si + 3HCl \rightarrow SiHCl_3 + H_2 \quad (4.2)$$

The gases are passed through a condenser and the resulting liquid is subjected to multiple fractional distillation to produce semiconductor grade $SiHCl_3$

(trichlorosilane). The $SiHCl_3$ is reduced by hydrogen by heating mixtures of the gases. This results in deposition of silicon in polycrystalline form on an electrically heated rod.

$$SiHCl_3 + H_2 \rightarrow Si + 3HCl \tag{4.3}$$

This step of producing polycrystalline silicon increases the production cost and needs improvement in technology for cost reduction.

4.3.3 *Polycrystalline Silicon to Single Crystal Silicon Wafers*

For fabrication of semiconductor devices, silicon must be both pure and be in a crystalline form with minimum number of defects in crystal structure. This is achieved by a commercial process often referred to as Czochralski process.

The polysilicon is melted in a crucible after addition of trace levels of one of the dopants. This is generally boron, a *p* type dopant, for silicon solar cells. With good temperature control, using a seed crystal, a large single crystal of silicon is pulled out from the melt in cylindrical form called ingots. The *n* or *p* type single crystal ingots are now commercially available.

The crystal ingot is sliced into thin wafers of about 300–500 micron thickness. This process results in more than 50% waste as cutting loss.

4.3.4 *Fabrication of Silicon Solar Cells*

Silicon slices, cut parallel to (1,1,1) plane from 1 ohm-cm *n* or *p* type single crystal ingots, are used in the fabrication of solar cells, for example Gangadhar et al. (1967), on the laboratory scale fabrication of silicon solar cells in India. These cells are now commercially manufactured. In what follows is a discussion based on Gangadhar et al. (1967).

The thickness of the wafers is 0.5 mm which is adequate for absorption of majority of sunlight useful in photovoltaic energy conversion. The shorter wavelengths of solar radiation are absorbed in the surface region of the wafer whereas the longer wavelengths penetrate deeper. The lifetime of minority carriers is then measured. Silicon wafers in which minority carrier lifetime is more than five microseconds are only used. The minority carrier lifetime is measured by the well known method of conductivity modulation in a point contact. If the minority carrier lifetime is low the charges generated by photons away from the junction are lost by recombination before reaching the junction.

The slices are lapped and polished to mirror finish on both the surfaces. This is followed by a 4–5 min etching in a mixture containing six parts of fuming nitric acid and one part of hydrofluoric acid. By keeping etchant temperature at about 15 °C and employing mechanical agitation, localized etching is minimized and a reproducible and excellent surface finish is achieved. The slices are rinsed in deionized water and degreased in a trilyne vapour degreaser designed for the purpose. The slices are preserved in acetone till diffusion. Fabrication of solar cells is carried out either by diffusing phosphorous on *p* type wafer or by diffusing Boron on *n* type wafer.

4.3.4.1 Diffusion of Phosphorus

Phosphorous is diffused in *p* type silicon slices using P_2O_5 as the source, using dry oxygen as carrier gas. Three specimens were used in each run. It consists of a silica tube passing through two controlled temperature zones with such temperature gradient between them as not to let the impurity condense in it. P_2O_5 was placed in the lower temperature zone, at a temperature of 220–250 °C determined by the volatility of the impurity. The vapours of P_2O_5 were carried with the help of dry oxygen over silicon slices maintained at some fixed temperature between 950 and 1020 °C (with the variation in temperature not exceeding ±2 °C). The slices were kept vertical in the boat to allow P_2O_5 vapours to react uniformly with both the surfaces. After about 5 min the boat carrying P_2O_5 was withdrawn and the oxygen flow continued till no more vapours were seen at the exhaust. Phosphorus was diffused in silicon slices for 45–60 min. The slices were then shifted to a temperature zone of 700 °C and were allowed to cool gradually.

4.3.4.2 Diffusion of Boron

In this process, boric acid (H_3BO_3) was used as an impurity source in *n* type silicon material. Boric acid was placed in the vicinity of the slices, maintained at a fixed temperature between 1000 and 1150 °C (with variation in temperature not exceeding ±2 °C). Nitrogen was used as the carrier gas and the quantity of moisture in air was critically controlled. It was essential to protect the surface from damage and to control the rate of diffusion of boron in silicon. The source material was removed from the diffusion zone after 5 min. Impurity was allowed to diffuse for 45–60 min which was followed by cooling of slices in a way similar to that during phosphorus diffusion.

4.3.5 *Electrical Contacts and Encapsulation*

In order to remove any oxide layer formed on the surface, slices obtained by either of the above mentioned processes were etched for 5 min in 48% hydrofluoric acid. This was followed by thorough rinsing in deionized water. The junction at the rim and the back of these slices was removed by lapping. The active surface was coated with some photo resistant paint leaving out gridded structure of optimum size; rim was fully covered with the paint. In the literature, the use of Kodak Photo Resist (KPR) is extensively recommended for the above purpose. However, KPR is an imported material. In India, the use of an alternate material known as Haunsa Resist/SS is recommended. This paint is not affected by acids and plating solutions and is readily soluble in acetone. The constituents of nickel plating solution are (gms/litre): nickel chloride, 30; sodium hypophosphite, 10 and ammonium chloride, 50.

After nickel deposition the slices were washed and Haunsa Resist/SS was cleaned off the surface. Wire connections were made using a special composition Wood's metal and tin with zinc chloride as flux. The cell was then encapsulated such that the top surface of the container helped in concentrating the sun's energy on the cell, thus increasing the effective output power.

The rectangular solar cells may have a line contact parallel to the longer edge. Such contact geometry results in increased probability of recombination for the charge carriers at larger distances from the contact. To improve collection of carriers, more interconnected contacts (grids) are provided on the active surface of the cell. Such a grid pattern leads to a lower contact resistance. The efficiency of such a cell depends on the width of the grid lines and their separation. While the grid lines reduce series resistance, they also reduce the active area available for incident light, thereby reducing the output power. Narrower grid lines reduce useful output power. If the separation of grid lines is too much, charge carriers would be lost by recombination, reducing the output power again. In practice, a compromise is made between these factors and an optimum spacing S' and width T'' of the grid lines are calculated. Knowing the terminal voltage V near the maximum power point and the other data of the solar cell without grid structure, the values of grid structure can be calculated from the equations for the grid structure (Wolf 1960)

$$T = 2^{5/4}\frac{\rho_T^{3/4}}{\rho_p^{1/2}}\left(BCj_0e^{BV}\right)^{1/2}W^{3/2} \tag{4.4}$$

$$S = \sqrt[3]{\frac{2T}{BC\rho_p j_0 e^{BV}} - \frac{2T}{3}} \tag{4.5}$$

where $C = 1 - \left(\frac{I_0}{I_L}\right)(e^{BV} - 1), B = \frac{q}{AKT}, \rho_T =$ the sheet resistance of the contact strip, ρ_p=the sheet resistance of the diffused p layer, $j_0 =$ the dark saturation current density, $I_L =$ the light generator current and $I_0 =$ the saturation current.

It is apparent from Eqs 4.4 and 4.5 that for higher light intensities the width of the grid lines should increase and their separation should decrease. Solar cells have highly polished surfaces and so the contacts made by electrodeless deposition of nickel or vacuum evaporation of gold do not adhere properly to the cell. The adherence of contacts is improved by sintering to form an alloy with silicon.

During sintering, the junction is penetrated, seriously degrading the cell characteristics. Silver-titanium contacts are used to avoid junction penetration (Smith et al. 1963). Silver-cerium contacts have also been developed for super blue cells.

About 30% of the incident light is reflected from the surface of silicon solar cells which have highly polished surfaces. Antireflection coatings are an obvious solution to reduce the reflection losses of useful light from the cell surface. Silicon monoxide has the desired properties for use as antireflection coating and it serves excellently well on bare silicon cells. An antireflection film of SiO is vacuum evaporated on to the surface of the solar cell. The quality of the film is checked by looking at the interference colours in diffused light after suspending these cells for 30 min in steam over boiling water. Adherence of the film is tested by the scotch tape test (Smith et al. 1963).

Because of the negative temperature coefficient for power output of the silicon solar cells, it would be desirable to keep the temperature of the cell as low as possible. Bare cells have an emittance of about 0.32 at room temperature, whereas for borosilicate glass it is 0.85. Thus, with a borosilicate glass slide kept in thermal contact with the cell the effective emittance of the cell would increase appreciably and the rise of cell temperature would not be as high. The use of a cover glass on uncoated cell leads to a slight improvement in cell performance.

4.3.6 Improved Solar Cell Structures and Techniques

Several solar cell materials and fabrication steps have been researched for improvement in conversion efficiency. The conversion efficiency of laboratory solar cells has increased from 17.1% (Milstein and Tsno 1984) to 24% (Wang et al. 1990). The production cells have always been approximately 5% less efficient (in absolute efficiency) than laboratory cells. The efficiency difference is mainly due to the difference in area of wafers used in production (50 cm^2) and laboratory cells (4 cm ^{2}). Corresponding metallization coverages are 10–15% and 2–3% and hence the difference in efficiencies. Most of the cell efficiency improvements have originated from improved cell structures and processing techniques. Initial improvement was obtained by reducing the recombination along the top cell surface using thermal oxide passivation; the improvements were augmented by the reduction in bulk recombination resulting from an alloyed aluminum rear contact. The structure may be referred to as Passivated Emitter Solar Cell (PESC) (Blakers and Green 1986). Further improvements in cell performance were obtained in the new cell structure, the Passivated Emitter and Rear Cell (PERC) (Blakers et al. 1989; Green et al. 1990). The main structural difference arises at the rear contact wherein a thermally grown oxide is used to electronically passivate most of the rear cell surface. The rear ohmic connection is made through the holes in passivating oxide. However, the resultant aluminum-silicon contact not only has a high contact resistance but also exhibits rectifying properties. The deficiency has been overcome by locally diffusing boron into contact areas at the rear of the cells. The resultant structure is referred to as Passivated Emitter Rear Locally (PERL) diffused cell (Wang et al. 1990). It exhibits 24% efficiency. The structure incorporates reasonably effective light trapping schemes. There seems to be some scope of improvement in this area.

The technology of fabrication, comprising large number of steps starting from quartzite to encapsulated silicon solar cells, had also undergone several developments over years. Spacing and width of contact grid lines are optimized to

(i) increase active area for incident light,
(ii) reduce contact resistance and
(iii) improve collection of charge carriers.

Antireflection coatings are employed to reduce reflection losses of incident light from the cell surface. Single crystal silicon solar cells have a higher conversion efficiency compared to the less expensive polycrystalline and amorphous cells.

Other technological improvements include better cell structures and processing techniques. Passivated emitter solar cell, passivated emitter and rear cell, and passivated emitter rear locally diffused cell have resulted from these improvements exhibiting a conversion efficiency of around 24%.

4.4 Indian Status

In India, mono-crystalline silicon solar cell was fabricated on laboratory scale at Indian Institute of Technology Delhi during 1967 (Jain et al. 1968). The cell active area was 2.4 cm^2 and the illumination level was 80 mW/cm^2. The conversion efficiency was around 10%. Solar cell fabrication has since reached a commercial status and several manufacturers are now marketing single crystal solar cells. Some of the physical and material characteristics of solar cell manufactured by Central Electronic Ltd. are as follows:

A.	Material:	Single crystal silicon
	Resistivity range:	0.5–3.0 ohm-cm
	Orientation:	$\langle 1, 0, 0 \rangle$
	Type:	*p*-type boron doped
	Thickness:	400 μm
	Cell area:	78.54 cm^2
B.	Process	
	Surface preparation:	Chemical etching 20–30 μm silicon
	Texturisation:	Low conc., alkaline solution by preferential etching

C. Diffusion

Gaseous diffusion:	Phosphorus ($POCl_3$)
Junction depth:	0.5 μm
Surface concentration:	10^{20} atoms/cm^2 graded
Temperature of diffusion:	900 °C

D. Front screen printing: Silver paste + drying

E. Back screen printing: Silver + Aluminum paste + drying + firing at 700 °C

Chapter 5
Mathematical Model of Transport Processes

5.1 Physical Model of Solar Cells

The physical processes that determine the role of a solar cell as an energy converter include absorption of light leading to generation of electron-hole pairs, separation of charges in depletion region, surface and bulk recombination and transport of charge carriers by diffusion and drift. Over the years efforts have been made to understand these processes (e.g. Fahrenbruch and Bube 1983; Green 1982) and ameliorate the efficiency of solar cells.

5.1.1 Excess Carrier Generation: Photo Generation

It is possible to generate excess carriers in the *p*- and *n*-regions by illuminating them with light. A flux of monochromatic light I_f (Watts/cm^2) incident on the surface of the *p*-region is considered. If the wavelength of light is λ (cm), the number of photons F_0' (photons/cm^2 sec) striking the surface will be given by

$$F_0' = \frac{I_f}{h\nu} = \frac{\lambda I_f}{hc} \tag{5.1}$$

where h is Planck's constant, and ν and c are frequency and velocity of light, respectively.

Part of this light is reflected from the surface and the remaining is transmitted. If the reflection coefficient of the surface is R, the transmitted flux is

$$F_0 = F_0'(1 - R) \tag{5.2}$$

N. D. Kaushika et al., *Solar Photovoltaics*,
https://doi.org/10.1007/978-3-319-72404-1_5

These photons are absorbed in accordance with Lambert's law which states that the absorption of light decays exponentially with distance into the material. The flux at a distance x from the surface is given by,

$$F = F_0 e^{(-\alpha x)} \text{photons} \tag{5.3}$$

where α is the absorption coefficient of light of wavelength λ in the semiconductor. The plot of α vs λ for silicon obtained from the available data is given in Fig. 5.1.

A volume of the plane bounded by x and $x + \Delta x$ (Wolf 1969) is considered. Assuming the quantum yield to be unity (i.e., an electron-hole pair is created for each photon absorbed), the number of electron-hole pairs per cm^3 per sec, injected between the planes at x and $x + \Delta x$ is given by

$$\frac{F_0 \left[e^{(-\alpha x)} - e^{(-\alpha (x+\Delta x))} \right]}{\Delta x} \tag{5.4}$$

Making Δx approach zero, the rate of generation of electron-hole pairs at a distance x from the surface is given by

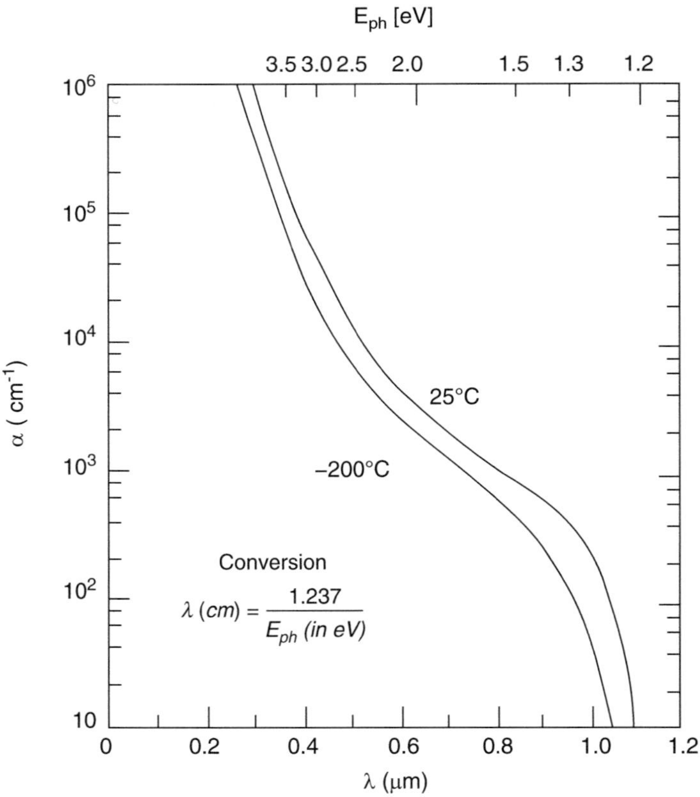

Fig. 5.1 Absorption coefficient (α) vs wavelength (λ) in single crystalline silicon (Wolf 1969)

$$g = F_0 \alpha \ e^{(-\alpha x)} \left(\text{pairs}/\text{cm}^3 \ \text{sec}\right) \tag{5.5}$$

The photons incident on the cell can be subdivided into five categories as described below:

(i) A portion of the incident light is reflected away. Some of the photons absorbed near the surface generate electron-hole pairs which recombine and do not reach the junction.
(ii) Photons absorbed near the junction, in both *p*- and *n*-regions, generating electron-hole pairs which are separated by the built-in field near the junction give rise to useful power, contributing to the conversion efficiency of the cell.
(iii) Photons absorbed deep in the collector region generate electron-hole pairs, so far from the junction that they recombine and, hence, are not useful.
(iv) Photons with insufficient energies to generate electron–hole pairs are absorbed and generate heat.
(v) Finally, some of the photons having longer wavelengths are not absorbed in either base or collector.

Furthermore, if the cell thickness is less, some of the photons which can generate electron-hole pairs are not absorbed, causing a lower output current. It is found that 0.4–0.5 mm thickness of silicon is adequate to absorb almost all the incident solar photons that generate electron-hole pairs which contribute to the power output of the cell. Fabrication of cells of greater thickness will only increase the number of photons absorbed away from the junction, which would not contribute to the power output of the cell. Also, longer wavelengths which have insufficient energy to generate electron-hole pairs are absorbed and generate heat, thereby increasing the cell temperature which in turn leads to a drop in cell power output (0.45% per degree Kelvin).

5.1.2 Built-in Field

The diffused *p-n* junction solar cell (Fig. 5.2) is characterized by a non-homogeneously doped (assume *p*-type base) over a constant impurity background of opposite conductivity referred to as collector. The non-homogeneous doping in the base region is the result of the fabrication of solar cell by impurity diffusion. The *p-n* junction solar cell under consideration is formed by the diffusion of *p*-type (acceptor) impurities into an *n*-type wafer containing *n*-type (donor) impurities. The impurity profile in the *p*-region is described by (i) a Gaussian function if the diffusion is carried out from an infinitesimally thin layer of dopant and (ii) a complementary error function, when the diffusing species is made available to the surface to maintain the concentration at a constant value (Jain and Al-Rifai 1967). It may be mentioned that the surface concentration of impurity at

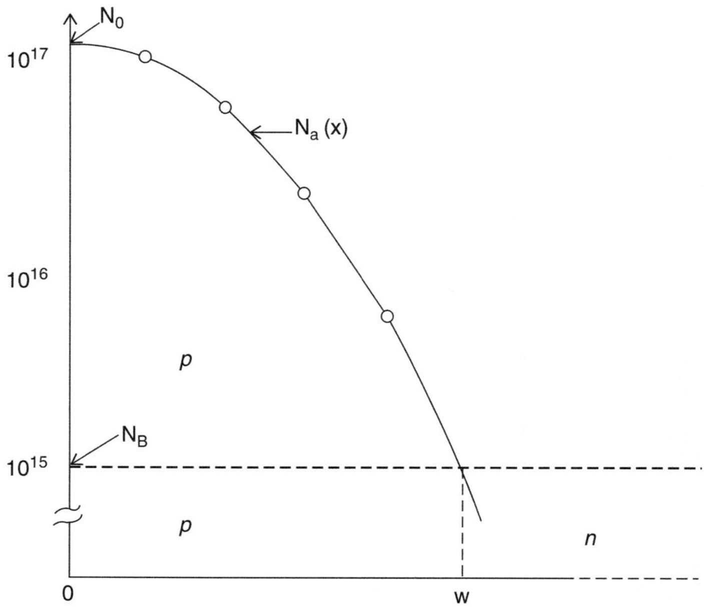

Fig. 5.2 Diffused *p-n* junction solar cell

the end of diffusion is determined by the time and temperature of diffusion, in the former cases while in the latter case it is limited by the solubility of the diffusion impurity at the temperature at which diffusion is carried out.

The resultant impurity profile results in a built-in electric field. This field influences the carrier transport in the diffused base region and hence it is to be considered in carrier transport process in this region. It is a common practice to approximate the impurity profile in the diffused region by a negative exponential. Such an approximation leads to a constant built-in field in place of a spatially varying built-in field resulting from a realistic profile described by complementary error function or Gaussian function. The exponential approximation of a constant built-in field in base region results in considerable simplification in the mathematical analysis of the physical processes (Gangadhar and Bhattacharya 1968; Fahrenbruch and Bube 1983) for photo diodes and solar cells. The expressions for built-in field for various impurity profiles can be derived as described below. The concentration in the diffused layer of the solar cell, arise from N_0 at the surface to N_B at the junction at $x = w$ as shown in Fig. 5.3.

The width of the *p*-region, over which there is a grading of the impurities, is designated by *w* (cm). The implications of the non-homogeneous impurity distribution will be considered presently. For the following discussions it is assumed that

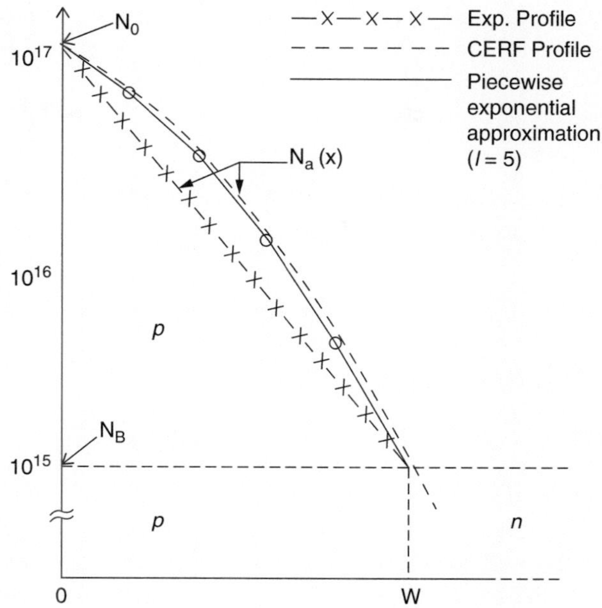

Fig. 5.3 Impurity profile and piecewise exponential approximation

(i) The material is non-degenerate and the Einstein relationship

$$D = \frac{kT}{q}\mu \tag{5.6}$$

is valid. D and μ are the diffusion coefficient and mobility of charge carriers respectively.

(ii) The impurities are completely ionized at room temperature.

Hence

$$p(x) = N_a(x)$$

where p and N_a are position dependent concentrations of holes and the acceptor impurities.

Due to the non-homogeneous impurity distribution and the consequent majority carrier distribution, by Fick's law, there is a natural tendency for diffusion. The diffusion current is given by

$$qD_p = \frac{dp(x)}{d(x)} \tag{5.7}$$

where D_p and $dp(x)/dx$ represent the diffusion constant and the concentration gradient of holes, respectively. This diffusion current is, at equilibrium, balanced by the drift current which is given by

$$-q\mu_p Ep(x) \tag{5.8}$$

when μ_P is the hole mobility and E (V/cm) is the built-in field, assumed positive from surface to the junction, in the p-region. Equating (5.7) and (5.8), we have

$$E(x) = \frac{D_P}{\mu_P}\frac{1}{p(x)}\frac{dp(x)}{dx} \tag{5.9}$$

and in view of Eq. (5.6).

$$E(x) = \frac{kT}{q}\frac{1}{p(x)}\frac{dp(x)}{dx} \tag{5.10}$$

Equation (5.10) shows that the built-in field is determined by the impurity distribution in the p-region. It is of practical importance to find the built-in field for different impurity distributions encountered in practice.

(i) *Gaussian impurity profile*

For the impurity profile of the form:

$$N_a(x) = N_0 e^{\frac{-x^2}{4D^*t}} \tag{5.11}$$

where D^* is the diffusion constant (cm^2 sec^{-1}) determined by the impurity and the temperature of diffusion which is carried out for a period of t seconds. N_0 and $N_a(x)$ represent the impurity concentrations (cm^3) at the surface and at any point x in the diffused p-region.

The built-in field, in view of Eqs (5.10) and (5.11), is given by

$$E(x) = -\frac{kT}{q}\frac{x}{2D^*t} \tag{5.12}$$

(ii) *Complementary error function profile*

For the impurity profile of the form:

$$N_a(x) = N_0 \operatorname{erfc}\left(\frac{x}{2\sqrt{D^*t}}\right) \tag{5.13}$$

The built-in field, in view of Eqs (5.10) and (5.13), is given by

$$E(x) = \frac{kT}{q}\frac{1}{\sqrt{\pi D^* t}}\frac{e^{\frac{-x^2}{4D^* t}}}{\operatorname{erfc}\left(\frac{x}{2\sqrt{D^* t}}\right)} \quad (5.14)$$

(iii) *Exponential profile*

For an impurity profile of the form:

$$N_a(x) = N_0 e^{\left(\frac{x}{2\sqrt{D^* t}}\right)} \quad (5.15)$$

The built-in field, in view of (5.10) and (5.15), is given by

$$E = -\frac{kT}{q}\frac{1}{N_a(x)}\frac{dN_a(x)}{dx} = -\frac{kT}{q}\frac{1}{2\sqrt{D^* t}} \quad (5.16)$$

The negative sign in Eqs (5.12), (5.14) and (5.16) shows that the field (assumed positive) is directed from junction to the surface, i.e., in a direction favourable for the minority carrier flow towards the junction. It is noted that the built-in field given by Eqs (5.12) and (5.14) is a function of distance into the diffused p-region. For an assumed exponential distribution, the built-in field given by Eq. (5.16) is constant for the whole of the p-region. Assuming $N_B = 10^{16}$ cm^{-3}, $N_0 = 10^{18}$ cm^{-3} and $w = 0.5$ μm, the built-in fields given by Eqs (5.12), (5.14) and (5.16) are calculated. Figure 5.4 shows the field and its variation in the diffused region for the above distributions. It is seen that for a Gaussian profile, the field is found to have a linear variation, while for the complementary error function (CERF) distribution, the field could be linearized incurring an error within 5%. It is also seen that (i) built-in field due to a Gaussian impurity profile has a zero value at the surface, (ii) the Gaussian profile has a larger field gradient compared to the corresponding CERF case, (iii) the field due to Gaussian profile being lower near the surface, increases rapidly towards the junction and exceeds the field due to the later at and near the junction than that due to the CERF distribution, and (iv) the constant field due to an assumed exponential profile has, in the diffused layer, field values greater near the surface, nearly equal midway and less than, near the junction, compared to the corresponding values resulting from the other two impurity profiles considered.

5.2 The Continuity Equation

The one dimensional continuity equation for electrons in the p-region of the p-n junction solar cell can be written as (Gangadhar et al. 1967).

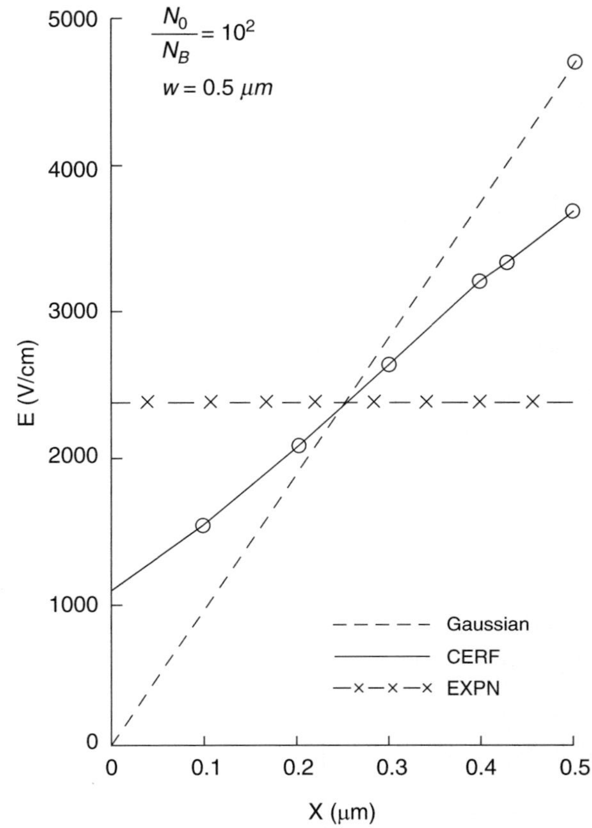

Fig. 5.4 Field and its variation in the diffused region

$$D_n \frac{d^2 n}{dx^2} + \mu_n E \frac{dn}{dx} - \frac{(n - n_{px})}{\tau_n} + F_0 \alpha e^{(-\alpha x)} = 0 \tag{5.17}$$

where D_n, μ_n and t_n are the average values of the diffusion coefficient, mobility and lifetime of electrons; $n_{px} = n_i^2/N(x)$, and n_i, n and n_{px} are the intrinsic, total and equilibrium concentration of electrons (cm^{-3}), respectively, in the p-region. Due to non-homogeneous doping, the mobility which depends on the value of impurity concentration becomes position dependent in the diffused p-region. The variation of mobility with impurity concentration (doping) is shown in Fig. 5.5 (Wolf 1969). Also, E, the built-in field has a spatial variation, as seen from Eqs (5.7) and (5.9). A general analytical solution of Eq. (5.17) for an arbitrary impurity profile taking both mobility and built-in field variations into account is difficult.

The methods described in the literature consist of those which consider, for the sake of mathematical simplicity, the impurity distribution in the diffused region to the exponential resulting in a single built-in field value and also employ a suitably averaged value of mobility (Jordan and Milnes 1960; Wolf 1963; Gangadhar and Kaushika 1992a). In what follows, the different methods for solving the continuity equation are reviewed.

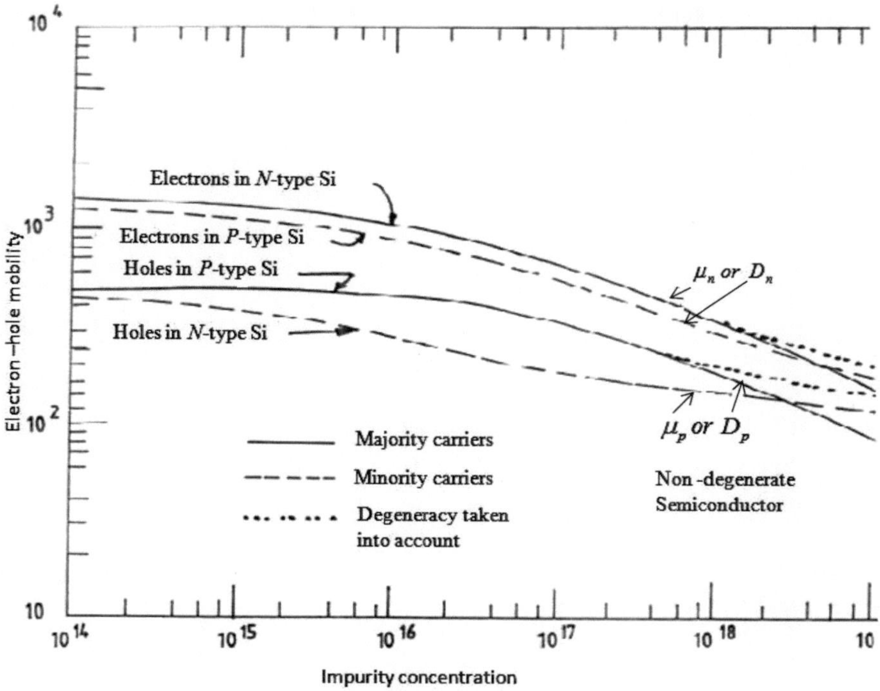

Fig. 5.5 Electron and hole mobility vs impurity concentration in silicon (Wolf 1969)

5.3 Solution of Continuity Equation

In constant field method/single exponential approximation method, the impurity distribution in the diffused region is assumed to be exponential, giving rise to a constant built-in field in it and its value is determined by the surface to background impurity concentration ratio and the width of the diffused region. It is also assumed that the mobility of minority carriers can be represented by a suitably averaged value for the whole of the non-homogeneously doped *p*-region. The *p*-region is divided into a number of sections and the mobility is averaged in each section. The average mobility for the whole of the *p*-region is obtained by taking the arithmetic mean of these averaged values. It is noted that the division of the *p*-region into about 20 parts gives a reasonably accurate estimate of the average mobility. The single exponential model is only an approximate one. This is because the practical impurity profiles described by a complementary error function of a Gaussian function give rise to a spatially varying built-in field. Also, the use of a single averaged mobility value for charge carriers does not adequately represent the practical situation wherein the mobility is position dependant.

The *p-n* junction solar cell is considered to have a thin diffused *p*-type base with a exponential acceptor impurity profile and a long *n*-type collector having uniform donor concentration. The governing one-dimensional continuity equation for the minority carriers in the *p*-type region, with a photon flux F'_0 of monochromatic light incident on the front surface of the cell is given by Gangadhar and Bhattacharya (1968):

$$D_n \frac{d^2 n}{dx^2} + \mu_n E \frac{dn}{dx} - \frac{n}{\tau_n} = -F_0 \alpha e^{(-\alpha x)} - \frac{n_{px}}{\tau_n} \tag{5.18}$$

where D_n, μ_n and t_n are the average values of the diffusion constant, mobility and lifetime of electrons, respectively; n and n_{px} are total and equilibrium concentration of electrons; α, the absorption coefficient of the semiconductor and E built-in electric field. Also, the acceptor profile (Figs. 5.2 and 5.3) will be taken as an exponential decaying from the surface, i.e.

$$N_a(x) = N_0 e^{-\beta x}$$

where N_0 is the impurity concentration at the surface, and β is a constant(cm^{-1}). From Eq. 5.11 we have,

$$E = \frac{1}{N_a} \frac{dN_a}{dx} \tag{5.19}$$

$$E = -\frac{kT}{q} \beta \tag{5.20}$$

and from Eqs (5.5) and (5.20).

$$\beta = \frac{1}{2\sqrt{Dt}} \tag{5.21}$$

$$n_{px} = N_{p0} e^{\beta x}$$

$$F_0 = F'_0(1 - R) \tag{5.22}$$

where n_{px} and n_{p0} are the minority carrier concentration at any point in the bulk and at the surface respectively.

N_a is the concentration of the acceptor impurity and R is the reflection coefficient. A low illumination level is assumed wherein number of minority carriers injected by light is much smaller than the concentration of majority carriers. Thus, conductivity modulation effects are neglected.

The appropriate boundary conditions at the surface $x = 0$ and at the junction $x = w$ are obtained as described below.

If the contact to the base is a perfect ohmic contact, i.e., no minority carriers are injected or extracted from the base, the only minority carriers lost are those which recombine at the surface of the *p*-layer i.e.,

$$\sigma(n - n_{px})|_{x=0} = D_n \frac{dn}{dx}|_{x=0} + \mu_n E_n|_{x=0} \tag{5.23}$$

and

$$n|_{x=w} = n_{pw} \tag{5.24}$$

where σ is the surface recombination velocity.

The solution of Eq. (5.18) is given by:

$$n = Ae^{m_1 x} + Be^{m_2 x} - \frac{\alpha F_0 e^{-\alpha x}}{D_n\left(\alpha^2 + \beta\alpha - \frac{1}{L_n^2}\right)} + n_{px} \tag{5.25}$$

where.

$$m = \frac{\beta \pm \sqrt{\left(\beta^2 + \frac{4}{L_n^2}\right)}}{2} \tag{5.26}$$

$$A = \frac{\alpha F_0}{D_n\left(\alpha^2 + \beta\alpha - \frac{1}{L_n^2}\right)} \times \left[(\sigma + D_n\alpha + D_n\beta)e^{m_2 w} - (\sigma + D_n m_2)e^{-\alpha w}\right] \times \frac{1}{\left[(\sigma + D_n m_2)e^{m_2 w} - (\sigma + D_n m_1)e^{m_1 w}\right]} \tag{5.27}$$

and

$$B = \frac{\alpha F_0}{D_n\left(\alpha^2 + \beta\alpha - \frac{1}{L_n^2}\right)} \times \left[(\sigma + D_n\alpha + D_n\beta)e^{m_2 w} - (\sigma + D_n m_2)e^{-\alpha w}\right] \times \frac{1}{\left[(\sigma + D_n m_1)e^{m_1 w} - (\sigma + D_n m_2)e^{m_2 w}\right]} \tag{5.28}$$

where L_n is the average diffusion length of electrons. The electronic component of current is given by:

$$J_n = qD_n \frac{dn}{dx}\,|_{x=w} + q\mu_n E_n\,|_{x=w} \tag{5.29}$$

Evaluating

$$\frac{dn}{dx}\,|_{x=w} \text{and } n|_{x=w}$$

it can be shown that:

$$j_n = \frac{qF_0\alpha}{D_n\left(\alpha^2 + \beta\alpha - \frac{1}{L_n^2}\right)} \times \left[\frac{m_1[(\sigma + D_n\alpha + D_n\beta)e^{m_2w} - (\sigma + D_nm_1)e^{-\alpha w}]e^{m_1w}}{(\sigma + D_nm_2)e^{m_2w} - (\sigma + D_nm_1)e^{m_1w}}\right] - \left[\frac{m_2[(\sigma + D_n\alpha + D_n\beta)e^{m_1w} - (\sigma + D_nm_2)e^{-\alpha w}]e^{m_2w}}{(\sigma + D_nm_2)e^{m_2w} - (\sigma + D_nm_1)e^{m_1w}}\right] + \alpha e^{-\alpha w} \tag{5.30}$$

Obviously j_n can be written as:

$$j_n = j_{nphoto} + j_{sndark} \tag{5.31}$$

where j_{nphoto} represents the electron component of the photocurrent which is a function of the incident photon flux. j_{sndark} stands for electron component of the dark reverse saturation current of the cell.

The above expressions are useful for the investigation of the overall performance of the *p-n* junction solar cell as a solar energy converter. In the present book calculations have been carried out for the case of silicon cells.

The total photocurrent j_{photo} due to creation of electron-hole pairs by the incident light is given by

$$j_{photo} = j_{nphoto} + j_{pphoto}$$

where j_{pphoto} is due to light generated minority carriers of the *n*-region (collector). For short wavelengths, i.e. for larger values of α, the electron hole pair generation is limited to the *p*-type base region. In such a case, the photocurrent is due to electrons of the *p*-type base region only, and is given by.

$$j_{photo} = j_{nphoto}$$

Figure 5.6 shows the dependence of the collection efficiency, defined as j_{photo}/qF_0 on the surface recombination parameter $\sigma w/D_n$ for built-in field parameter $\beta w = 0.1$ and with $w/L_n = 0.2$. It may be mentioned that $\beta w = 0$ represents the case of a solar cell having a uniform concentration of *p*-type impurities in the base region.

It may be noted that for higher built-in field the rate of decrease of collection hhefficiency with respect to surface recombination $\sigma w/D_n$ is reduced. It means that under such a condition the performance of the solar cell is less sensitive to the changes of surface properties.

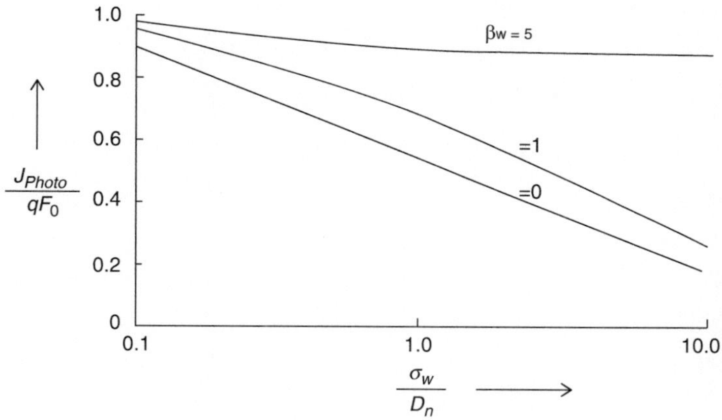

Fig. 5.6 Collection efficiency j_{photo}/qF_0 vs dimensionless surface recombination $\sigma w/D_n$ with dimensionless field βw as parameter

5.4 Piecewise Exponential Computational Scheme

This method utilizes the simplicity of exponential model and yet effectively takes into account the field and mobility gradients. The method had been developed earlier (Gangadhar and Bhattacharya 1968) for the study of admittance characteristics of the *p-n* junction diode with a narrow diffused region; subsequently for the study of spectral response of solar cells by Gangadhar and Kaushika (1992b). In this method, the diffused *p*- region is divided into a number of sections, say *l*, and in each section, the impurity profile $N = N(x)$ is approximated by a separate exponential such that

$$E_i = \frac{kT}{q}\beta_i = a \text{ constant} \tag{5.32}$$

$$\text{with}\,\beta_i = \frac{1}{w_i}\ln\frac{N_{i-1}}{N_i}, i = 1, 2, 3 \ldots l \tag{5.33}$$

where w_i is the width of the ith section and N_i is the actual impurity concentration at the interface of the ith and $(i + 1)^{\text{th}}$ sections. This procedure, where the three types of impurity profiles are subjected to a piecewise exponential approximation, is shown in Figs. 5.2 and 5.3. For the three types of profiles $N_0 = 10^{17}\ \text{cm}^{-3}$ and $N_B = 10^{15}\ \text{cm}^{-3}$, are assumed. The impurity concentration due to an assumed exponential distribution significantly differs from the Gaussian or complementary error function distribution.

It is found that

(i) for a large number of sections, i.e., for a large value of *l*, E_i as obtained from Eqs (5.7), (5.9) and (5.11) is close to the actual average field value in each section and

Table 5.1 Average field values in the diffused region

Impurity distribution	Average built-in field, E_i, V cm^{-1}, in different sections									
	(1)	(2)	(3)	(4)	(5)	(6)	(7)	(8)	(9)	(10)
Gaussian	120	350	590	820	1050	1290	1590	1760	1990	2210
CERF	582	730	870	1020	1160	1310	1450	1600	1745	1890
Exponential	1190	1190	1190	1190	1190	1190	1190	1190	1190	1190

$N_0 = 10^{17}$ cm^{-3}, $N_B = 10^{15}$ cm^{-3}, $w = 1$ μm

(ii) the division of the p-region into 10 sections, i.e., $l = 10$ serves the purpose with E_i, values approaching the average field in the ith section.

The E_i values with $w = 1$ μm, $N_0 = 10^{17}$ cm^{-3} and $N_B = 10^{15}$ cm^{-3} have been calculated for Gaussian, complementary error function and exponential impurity distributions and are shown in Table 5.1 (Gangadhar and Kaushika 1992b). Similarly the minority carrier mobility values μ_n, averaged in each of the l-sections denoted by μ_{ni} for the three impurity profiles can be calculated (Gangadhar and Kaushika, 1992b).

The method is applicable to all types of impurity profiles and using a few sections a reasonably accurate characterization of the transport property of an homogeneously doped device is possible.

The above approach of 'piecewise exponential approximation' had been made use of in the analysis of spectral response of solar cells in the book. This method has the following features:

(i) The diffused region is divided into a number of sections (Figs. 5.2 and 5.3) and the impurity distribution is approximated by an exponential in each of the sections. As can be seen from Fig. 5.3, this amounts to taking the actual impurity concentration unlike the single exponential approximation. The proximity to the actual value is more whenever the number of sections is increased.
(ii) The mobility of minority carriers is the averaged values in each of the sections, whereas a single averaged value is used in the single exponential approximation. Thus, the mobility value is nearer to the actual value especially for a large number of sections. This in a way takes care of mobility variation.
(iii) The diffusion coefficient approaches the actual value and its variation is taken into account as discussed above, for mobility.

In view of above, it is seen that the method of piecewise exponential approximation is a realistic approach as it takes into account the actual impurity concentration, its variation as well as the value and the variation of built-in field; values of mobility, its variation; and value of diffusion coefficient, and its variation; to a considerable extent. Also, the method is independent of whether the resultant impurity concentration in the diffused region, depending on the technique of fabrication, is described by an exponential, Gaussian, complementary error function or any other function.

Formulation

Let us consider a *p*-*n* junction solar cell having a thin diffused *p*-type base and long homogeneously doped *n*-type collector. The impurity concentration and its distribution in the base region gives rise to spatially varying built-in field and mobility values. The variations may be taken into account by a method known as integration by piecewise exponential approximation developed earlier for characterization of a narrow diffused diode (Bhattacharya and Gangadhar 1971). In this approach the diffused region is divided into a number of sections, say l, and in each section, the built-in field E_i and β_i may be expressed as follows:

(i) Uniform distribution: $\beta_i = 0$ $(i = 1, 2, \ldots, l)$
(ii) Exponential distribution: $\beta_i = \text{constant}$ $(i = 1, 2, 3\ldots, l)$
(iii) Gaussian and CERF distribution: β_i is variable with i.

A typical *p*-type Gaussian impurity profile in the base region and its piecewise exponential approximation are illustrated in Fig. 5.2. Figure 5.3 shows a complementary error function profile in the diffused *p*-region subjected to piecewise exponential approximation. The built-in field in each of the section is given by Eq. (5.32). For a large number of sections the built-in field E_i is close to average field in each section. The one dimensional continuity equation for electrons in the ith section of the diffused *p*-region, with a photon flux F_0' of monochromatic light incident on the front surface of the cell is given by

$$D_{ni}\frac{\partial^2 n_i}{\partial x^2} + \mu_{ni}E_i\frac{\partial n_i}{\partial x} - \frac{n_i - n_{pxi}}{\tau_n} + F_0\alpha e^{-\alpha x} = 0, i = 1, 2, 3\ldots\ldots, l \quad (5.34)$$

where D_{ni} and μ_{ni} are the diffusion coefficient and mobility suitably averaged in the ith section, τ_n is the lifetime of electrons, n_i and n_{pxi} are the total and equilibrium concentration of electrons and α is the absorption coefficient of the incident radiation in the semiconductor. By making a large number of sections, the diffusion coefficient and mobility values approach actual values and their spatial variation.

$$n_{pix} = \frac{n_i^2}{n_i|_x} \quad (5.35)$$

where n_i is the intrinsic carrier concentration, and.

$$F_0 = F_0'(1 - R) \quad (5.36)$$

where R is the reflection coefficient.

The solution of Eq. (5.34) is given by:

$$n_i = A_i e^{m_i x} + B_i e^{l_i x} - G_i F_0 e^{-\alpha x} + n_{pxi} \quad (5.37)$$

where.

$$m_i, l_i = \frac{\beta_i}{2} \pm \sqrt{\frac{\beta_i^2}{4} + \frac{1}{D_{ni}\tau_n}}$$

and

$$G_i = D_{ni} = \left[\alpha + \beta_i - \frac{1}{\alpha D_{ni}\tau_n}\right]^{-1} \tag{5.38}$$

The constants A_i ($i = 1, 2,\ldots, l$) and B_i ($i = 1, 2,\ldots, l$) are evaluated using the boundary conditions corresponding to each section. The assumptions made to obtain the boundary conditions are:

1. The minority carrier density, at the interface of any two sections, is continuous; which gives:

$$\begin{array}{c} A_i e^{m_i x_i} + B_i e^{l_i x_i} - A_{i+1} e^{m_{i+1} x_i} - B_{i+1} e^{l_{i+1} x_i} - F_0 e^{-\alpha x_i}(G_i - G_{i+1}) = 0 \\ i = 1, 2, \ldots (l-1) \end{array} \tag{5.39}$$

2. The minority carrier current density (due to diffusion and drift) is also continuous at the interface of any two adjacent sections. As a result,

$$\begin{array}{c} A_i l_i e^{m_i x_i} + B_i m_i e^{l x_i} - A_{i+1} l_{i+1} e^{m_{i+1} x_i} - B_{i+1} m_{i+1} e^{l_{i+1} x_i} \\ + D_{ni} G_i F_0 e^{-\alpha x_i}[\alpha + \beta_i] - D_{n(i+1)} G_{i+1} F_0 e^{-\alpha x_{i+1}}\left[\alpha + \beta_{i+1}\right] = 0 \\ i = 1, 2, \ldots (l-1) \end{array} \tag{5.40}$$

Equations (5.39) and (5.40) give ($2l - 2$) boundary conditions and two more are obtained by considering the conditions at the surface ($x = 0$) and at the junction $x = w$

$$\sigma(n) \mid_{x=0} = D_{n1} \frac{\partial n_1}{\partial x} \mid_{x=0} + \mu n 1 E_1 n_1 \mid_{x=0} \tag{5.41}$$

and

$$n_1 = n_{px} \mid_{x=w} \tag{5.42}$$

where σ is the recombination velocity at the *p*-type surface and w is the depth of the *p-n* junction. The constants A_i and B_i can be calculated using Eqs (5.39, 5.40, 5.41 and 5.42), through the Determinant method of solving simultaneous equations. The electron component of the photo current density is given by:

$$j_{nphoto} = qD_{n1} \frac{\partial n_1}{\partial x} \mid_{x=w} + q\mu_{n1} E_1 n_1 \mid_{x=w} \tag{5.43}$$

Finally, the quantum efficiency, η_Q and the collection efficiency, η_C of the diffused base region, are given by

$$\eta_Q = \frac{j_{nphoto}}{qF_0} \tag{5.44}$$

$$\eta_C = \frac{j_{nphoto}}{qF_0[1 - e^{-\alpha w}]} \tag{5.45}$$

5.5 Illustrative Results

The efficiencies of solar cell having a Gaussian distribution in the base region have been computed corresponding to solar radiation wavelength range of 0.4–0.8 μm. Data used are $N_B = 10^{15}$ cm^{-3}, $N_0 = 10^{17}$ cm^{-3}, $w = 1$ μm, $\tau_n = 10^{-9}$ sec and $1 = 10$.

Table 5.2 (Gangadhar and Kaushika 1992b) shows values of η_0 and η_C obtained for different wavelengths and their variation with recombination velocity at the surface.

With a view to see the variation of the spectral response with base region acceptor impurity profile and associated built-in field and mobility values, calculation similar to the above have been made for the four distributions namely complementary error function, Gaussian, exponential and epitaxial, and compared in Table 5.3 (Gangadhar and Kaushika 1992b).

It is seen that the efficiencies are different for different distributions. Hence, the realistic distribution must only be considered and must not always be approximated by a single exponential distribution. The present method enables the realistic distribution to be taken into account and is adaptable on a microcomputer.

Table 5.2 Spectral response (with σ as variable)

	o' (cm/s)							
	10		10^5		10^6		10^7	
λ (μm)	η_Q	η_C	η_Q	η_C	η_Q	η_C	η_Q	η_C
0.40	0.916	0.916	0.779	0.779	0.429	0.429	0.263	0.263
0.45	0.862	0.906	0.791	0.831	0.608	0.640	0.522	0.549
0.50	0.690	0.945	0.645	0.884	0.533	0.730	0.479	0.657
0.55	0.524	0.954	0.495	0.900	0.420	0.764	0.384	0.694
0.60	0.376	0.963	0.357	0.914	0.307	0.787	0.283	0.727
0.65	0.248	0.954	0.236	0.907	0.205	0.790	0.191	0.733
0.70	0.189	0.945	0.180	0.901	0.157	0.786	0.146	0.732
0.75	0.150	0.961	0.143	0.915	0.125	0.801	0.117	0.747
0.80	0.091	0.959	0.087	0.916	0.077	0.806	0.071	0.751

Table 5.3 Spectral response (with impurity profile as variable)

	CERF		Gaussian		Exponential		Epitaxial	
λ (μm)	η_Q	η_C	η_Q	η_C	η_Q	η_C	η_Q	η_C
0.40	0.931	0.931	0.916	0.916	0.927	0.927	0.826	0.826
0.45	0.899	0.946	0.862	0.906	0.890	0.936	0.804	0.846
0.50	0.695	0.953	0.690	0.945	0.685	0.941	0.628	0.863
0.55	0.528	0.961	0.524	0.954	0.517	0.939	0.479	0.870
0.60	0.378	0.970	0.376	0.963	0.367	0.933	0.344	0.874
0.65	0.250	0.957	0.248	0.954	0.237	0.917	0.227	0.876
0.70	0.190	0.951	0.189	0.945	0.176	0.893	0.173	0.878
0.75	0.151	0.965	0.150	0.961	0.134	0.854	0.137	0.879
0.80	0.092	0.965	0.091	0.959	0.022	0.235	0.084	0.879

Chapter 6
Electrical Characteristics of Solar Cells

6.1 Solar Cell: A Power Source

To understand the behaviour of a solar cell as an electric power source, let us review the familiar characteristics of *p-n* junction diode and its behaviour under forward and reverse bias when dark. When *n*-type silicon semiconductor (characterized by electrons as majority carriers and holes as minority carriers) is in metallurgical contact with a piece of *p*-type semiconductor (characterized by holes as majority carriers and electrons as minority carriers) the boundary formed is *p-n* junction (Fig. 6.1). Owing to the process of natural diffusion electrons would flow from regions of high concentration (*n*-type) to the regions of low concentration (*p*-type) and the similar diffusion for holes. The electrons leaving the *n*-type side will expose positive charges in *n*-type side. The holes leaving the *p*-type side will expose the negative charges. This will result in a charge imbalance on both sides; it will set up an electric field that will oppose the natural diffusion tendency of electrons and holes and an equilibrium situation will be obtained. The region of transition is known as depletion layer, because the region is depleted of holes and electrons. The diode behaviour of such a *p-n* junction in dark conditions for forward bias and reverse bias conditions is well known and is shown in Fig. 6.2.

A forward bias voltage (*p* connected to positive of the battery and *n* to the negative) reduces the field in depletion region allowing a diffusion of majority carriers across the junction to the region where they are in minority resulting the I-V curve of first quadrant in Fig. 6.2. In reverse bias condition, electric field in depletion region is increased which strongly opposes the diffusion of majority carriers to zero. However, the field direction and magnitude is such that it attracts minority carriers from the neutral regions and moves them by drift across the depletion region resulting in a negative or reverse current through the structure. The concentration of minority carriers is order of magnitude lower than the majority carriers for each of the two regions. Consequently the reverse current is orders of magnitude lower than the forward current. Thus a solar cell under dark conditions acts as a rectifier.

N. D. Kaushika et al., *Solar Photovoltaics*,
https://doi.org/10.1007/978-3-319-72404-1_6

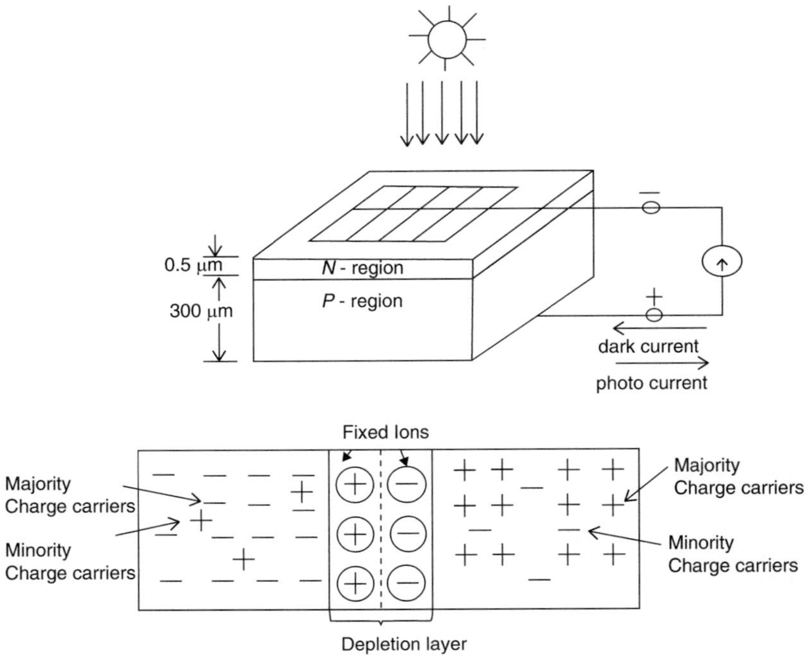

Fig. 6.1 *p-n* junction diode behaviour of solar cell

Fig. 6.2 Junction diode behaviour in dark condition for forward bias and reverse bias conditions

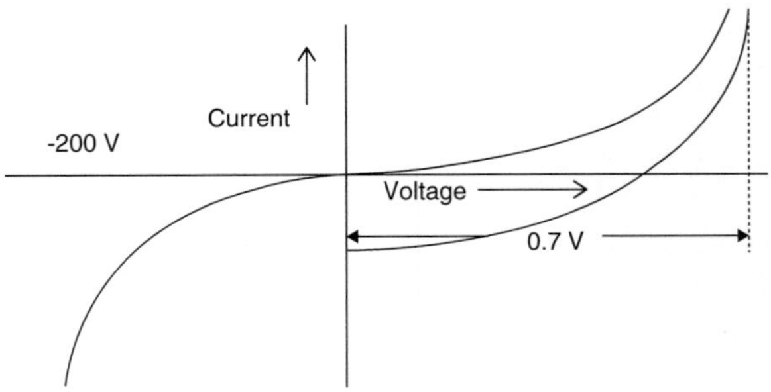

Under illumination semiconductor plasma is generated in the solar cell and it is separated by the drift of minority carriers across the depletion regions resulting in large reverse current. The light generated is superimposed onto the normal rectifying current-voltage characteristics of the diode. This results in the I-V curve covering the fourth quadrant where electrical power ($I_\nu V$) is negative which means that the power can be extracted from the device.

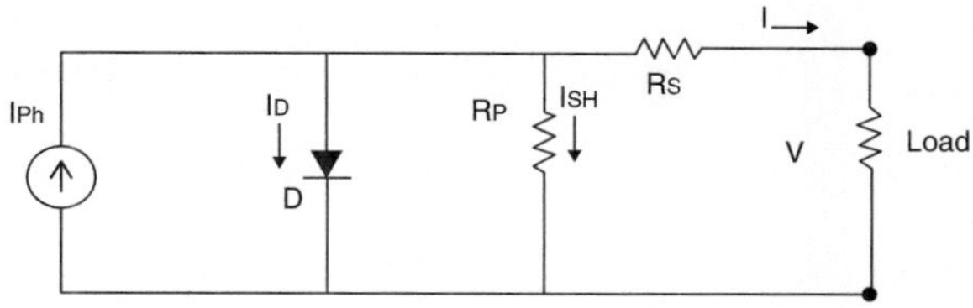

Fig. 6.3 Equivalent circuit of a solar cell

6.2 Equivalent Circuit

A solar cell may be regarded as a current source shunted by a diode and a resistance R_p, in series with R_s (lumped internal series resistance) to produce a terminal voltage V and current, I through the load resistance. This equivalent circuit is shown in Fig. 6.3.

It consists of a constant current generator shunted by the junction which acts like a positively biased diode. R_s is the lumped internal series resistance and R_L is the load resistance. The series resistance arises mainly due to

(i) the thin diffused layer on a silicon solar cell and
(ii) ohmic contact on the two opposite surfaces and bulk silicon resistivity.

The shunt resistance R_p is the result of the leakage current across the junction and the edge of the solar cell. Shunt resistance and capacitance can normally be neglected. Furthermore, in view of the fact that solar cell is a power source in contrast to an ordinary diode which absorbs power, in practice it is customary to regard the direction of photo-generated current to be positive.

6.3 Current-Voltage Characteristics

Following the above solar cell equivalent circuit I-V characteristics may be expressed as follows:

$$I = I_G - I_j$$
$$= I_G - I_0\left[e^{\left(\frac{q(V-IR_S)}{AKT}\right)} - 1\right] - \frac{V - IR_s}{R_p}$$

Normally we take $R_p \to \infty$ and $R_s \to 0$, which gives

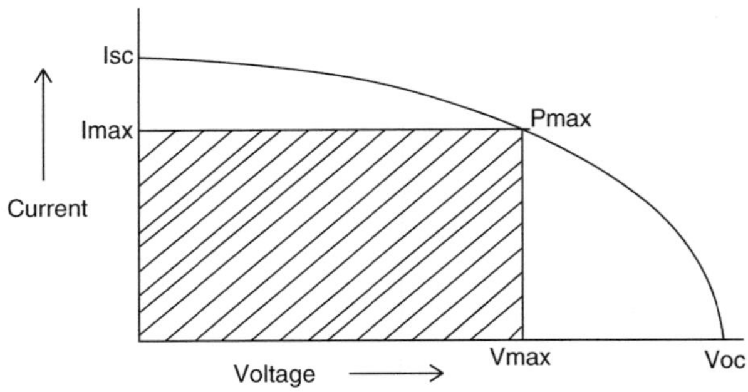

Fig. 6.4 I-V characteristics of crystalline silicon solar cell

$$I = I_G - I_0\left[e^{\left(\frac{qV}{AKT}\right)} - 1\right] \quad (6.1)$$

where I_G = photo-generated current, I_0 = dark saturation current of the diode; it is proportional to $e^{(-Eg/KT)}$, q = electronic charge (1.6022×10^{-19} C), K = Boltzmann constant (1.38×10^{23}Joule/°K), T = absolute temperature of cell and V = terminal voltage.

In accordance with Eq. 6.1 current-voltage curve of a crystalline silicon solar cell will be represented as in Fig. 6.4. The parameters indicated therein correspond to Standard Test Conditions (STC); cell temperature of 25 °C and solar irradiance of 1000 W/m^2.

The short circuit condition corresponds to zero voltage across the terminals of the cell; I_j is very small and practically all the generated current passes through the external circuit. The short circuit current is therefore a useful measure of photo-generated current and the two terms mean the same. The open circuit voltage (V_{oc}) corresponds to $I = 0$, $V = V_{oc}$ which from Eq. (6.1) is given as follows:

$$V_{oc} = \frac{AKT}{q}\log_a\left[\frac{I_G}{I_0} + 1\right] \quad (6.2)$$

Furthermore, solar cell current-voltage (I-V) relationship is such that the power output is zero at short circuit and open circuit and the output is maximum (P_{max}) at intermediate point (V_{max}, I_{max}). For maximum power output of cell, its load operation corresponds to V_{max}, I_{max} conditions. Maximum power, P_{max} corresponds to the area of largest rectangle that can be fitted in I-V curve. The conversion efficiency is the maximum power output expressed as a percentage of the input power. The cell conversion efficiency (η) is given by

$$\eta = \frac{P_{max}}{\text{Irradiance} \times \text{Area}} \times 100 \quad (6.3)$$

Obviously, nearer the I-V curve to rectangular shape, higher is the maximum power and conversion efficiency. The rectilinearity of I-V curve is represented by a parameter referred to as fill-factor which is defined as

$$ff = \frac{P_{max}}{I_{sc} \times V_{oc}} \tag{6.4}$$

6.4 Performance Measurement Procedures

The performance of a crystalline solar cell is a sensitive function of

(i) Solar irradiance,
(ii) Cell temperature and
(iii) Spectral irradiance distribution.

Consequently, the experimental arrangements for determining the performance of a solar PV device—be it a cell, a module, a sub array or an array—must involve the device exposure to known temperature and stable sunlight (natural or simulated).

The measurement of I-V characteristics must involve the simultaneous recording of the solar irradiance and cell temperature. For uniform interpretation it is often desired that the measured characteristics be transposed to Standard Test Conditions (STC) which are as follows:

1. Irradiance: 1000 W/m^2
2. Cell temperature: 25 °C
3. Reference spectral irradiance distribution due to Bird et al. (1983).

The power output at STC is often referred to as rated or peak power.

In natural sunlight the spectral content varies with location, weather, time and day of the year. Similarly in solar simulator it depends on light source and the optical system that is affected by aging. In practice it has been found that spectral changes can affect outdoor measurements on PV device as much as 15%; if the irradiance is measured with an instrument that is not spectrally selective such as pyranometer or other thermopile type radiometer. In such a situation the measurement on PV device cannot be related to a reference spectral irradiance distribution for rating purposes. This problem may be bypassed by measuring the irradiance with a reference device that has essentially the same spectral response as the test (PV) specimen and has been calibrated in terms of short circuit current per unit of irradiance of reference spectral distribution. The solar device measures the irradiance in terms of reference distribution and the specimen measurements give the response corresponding to same reference distribution. In this way, location, time and weather conditions in outdoor conditions and aging of simulator in indoors is automatically taken care. In experimental measurements we have studied I-V characteristics with known cell temperature (or else have affected temperature correction) and in stable solar irradiance which has been measured by a solarimeter (Surya Mapi) of same spectral response as PV (test) specimen.

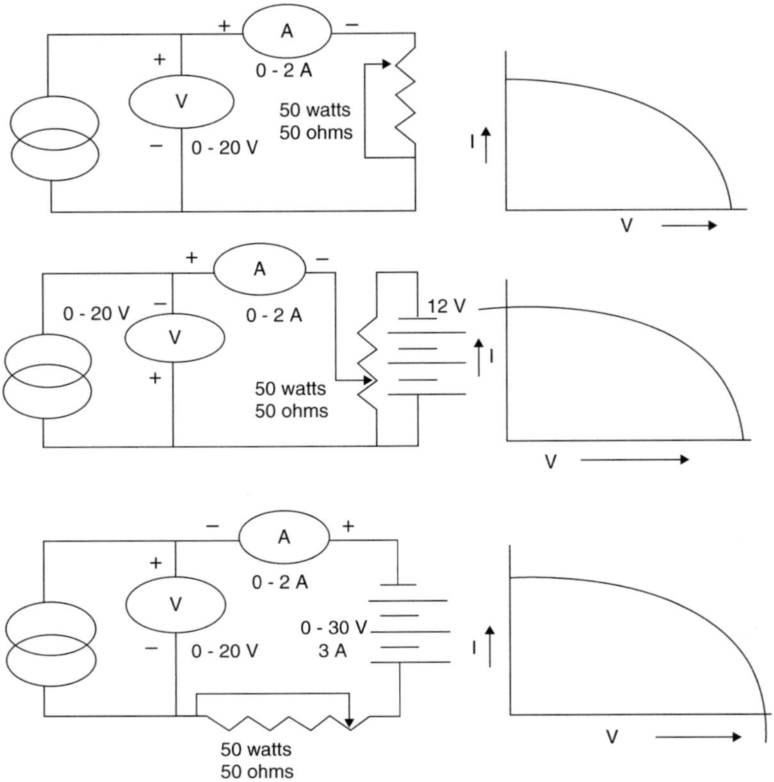

Fig. 6.5 Schematic diagram of measurement procedure of I-V characteristics

The general schematic of measurement procedure is illustrated in Fig. 6.5. The I-V curve can be traced by varying load through range. The variable load should preferably be an electronic circuit which sweeps the load from short circuit to open circuit condition. The accuracy of measurements is in conformity with international standard procedures as indicated below:

Parameter	Accuracy
Temperature	± 1 °C
Current and voltage	$\pm 0.5\%$

Furthermore, the open circuit voltage V_{oc} can be obtained by isolating the test specimen by means of a switch. The short circuit current in ideal situation is measured at zero voltage and electronic load (negative) is used to offset the voltage drop across the external series resistance. Yet another approach could be to measure the short circuit current at a voltage up to 3% of open circuit voltage and extrapolate it to zero voltage.

Our experiments have also involved the measurements of I-V characteristics in natural sunlight. To ensure stable irradiance, measurements were carried out on clear days and during 30 minutes of local noon hours.

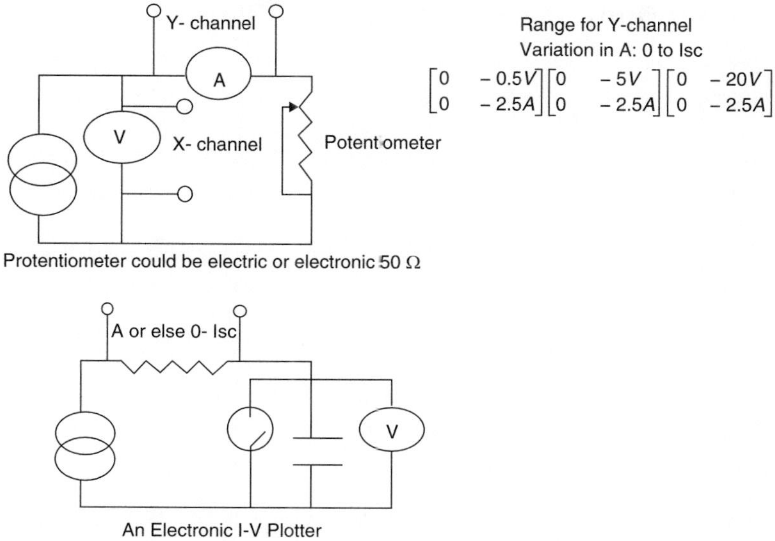

Fig. 6.6 Electronic measurement of I-V characteristics

For small power units (e.g. a solar cell or a small string/module of solar cells) manually adjusted rheostat may also be used. Typical values of variable load and precision resistors for such cases were 50 Ω (50 W) and (0.5–1) Ω respectively. In case of simple manual observational approach, the two circuits as shown in Fig. 6.6 may be used.

The power produced in each of the above cases is very small and there must be an intermediate value of R_L for which power is maximum, P_{max}. The corresponding output voltage and current are referred to as optimum voltage, V_m and optimum current, I_m, respectively.

$$P_{max} = I_m V_m \tag{6.5}$$

The electrical quality of a solar cell is measured in terms of fill factor, *ff*.

$$ff = \frac{V_m \ I_m}{V_{oc} \ I_{sc}} \tag{6.6}$$

Thus V_{oc}, I_{sc}, V_m, I_m, P_{max} and *ff* are the basic electrical parameters of solar cells which are of interest in their practical utilization.

To determine the I–V characteristics of a solar cell, the voltage across the cell is measured using a voltmeter and the current flowing through it is measured using an ammeter. The load resistance is varied and the corresponding voltage and current are measured at a particular solar insolation, measured using a solarimeter. A typical silicon solar cell was fabricated at the Indian Institute of Technology, Delhi during 1967. The cell active area was 2.4 cm^2 and the illumination level was 80 mW/cm^2. The conversion efficiency was around 10%.

6.5 Temperature and Solar Irradiance Effects

The I-V characteristics for a silicon solar cell fabricated at the Indian Institute of Technology, New Delhi (Jain et al. 1968) was measured at different temperatures as early as in 1968. The solar cell had been placed in a thermostat whose temperature is varied over a wide range of temperatures. The cell is kept under constant illumination by a lamp. The cell temperature had been measured with a thermocouple and a temperature indicator. The objective of the study was to measure the I-V characteristics of the solar cell, the short circuit current, open circuit voltage and power output of the cell at different temperatures. The study enables assessment of their temperature dependence and concludes that

(i) The short circuit current increases slightly with increase of temperature.
(ii) The open circuit voltage decreases rapidly at the rate of 2.3 mV $°C^{-1}$.
(iii) Power output of the cell decreases at the rate of about 0.6% $°C^{-1}$.

Subsequently various authors have investigated the effects of changes in irradiance and temperatures. The typical temperature coefficients (Treble 1991) are:

$$\frac{dI_{SC}}{dT} = 0.1\,\mathrm{Am}^{-2\circ}\mathrm{C}^{-1}$$
$$\frac{dV_{OC}}{dT} = -2.2\,\mathrm{mV}^{\circ}\mathrm{C}^{-1}$$
$$\frac{dP_{MAX}}{dT} = -0.5\%\ ^{\circ}\mathrm{C}^{-1}$$

With the increase in the intensity of illumination, the short circuit current increases linearly up to a reasonably high level of illumination. The open circuit voltage increases slightly with increase in the intensity of illumination. Also, the open circuit voltage is less sensitive to increase of illumination as well as variations in illumination level.

At higher levels of illumination, it is desirable to provide cooling to the cells. This is owing to the fact that at higher levels of illumination, the power output of the cell decreases with the increase of temperature.

Chapter 7
Solar PV Module and Array Network

7.1 Solar PV Module

The smallest operational unit of all SPV systems is the solar cell. Its size is chosen according to planned application. To obtain the desired voltage and current ratings the individual solar cells are connected in series and/or parallel. A string of cells connected in a pure series arrangement is referred to as series string or sub string. A group of series string or sub strings connected in parallel forms a series blocks. A branch circuit is composed of a series blocks between the positive and negative terminals of a power unit. An array is a collection of branch circuits (Fig. 7.1). The array is constructed in the field and transported from factory to field in small units called *modules*. The modules can be used to construct mW to MW size SPV arrays. Module is a panel for structural purposes having solar cells connected in series or parallel encapsulated in weather resistant laminate to protect the cells and enclosed in the sturdy corrosion resistant frame.

It may be noted that present-day modules are made from non-circular solar cells. Based upon the state and quality of silicon material three types of solar cells/modules have been produced and marketed. Mono-crystalline silicon solar cells have almost no defects and impurities. They have solar conversion efficiency of 15–17%. They are expensive to produce. Polycrystalline silicon cells/modules are slightly poorer grade monocrystalline silicon or semiconductor grade silicon. They exhibit white speckles on surface due to impurities. They are less expensive since simpler processes are involved in their production. They have solar conversion efficiency of 11–13%. Amorphous silicon cells/modules are made from amorphous silicon rather than silicon of crystal structure. They absorb light more effectively and are relatively thinner. They have lower conversion efficiency of 5–7%. However the reduced cost often overweighs reduced efficiency leading to a net increase in ratio of performance to cost.

Crystalline silicon modules are presently the dominant commercial product and deliver approximately 100–130 W/m^2 at STC. Amorphous silicon (a-Si) thin-film

N. D. Kaushika et al., *Solar Photovoltaics*,
https://doi.org/10.1007/978-3-319-72404-1_7

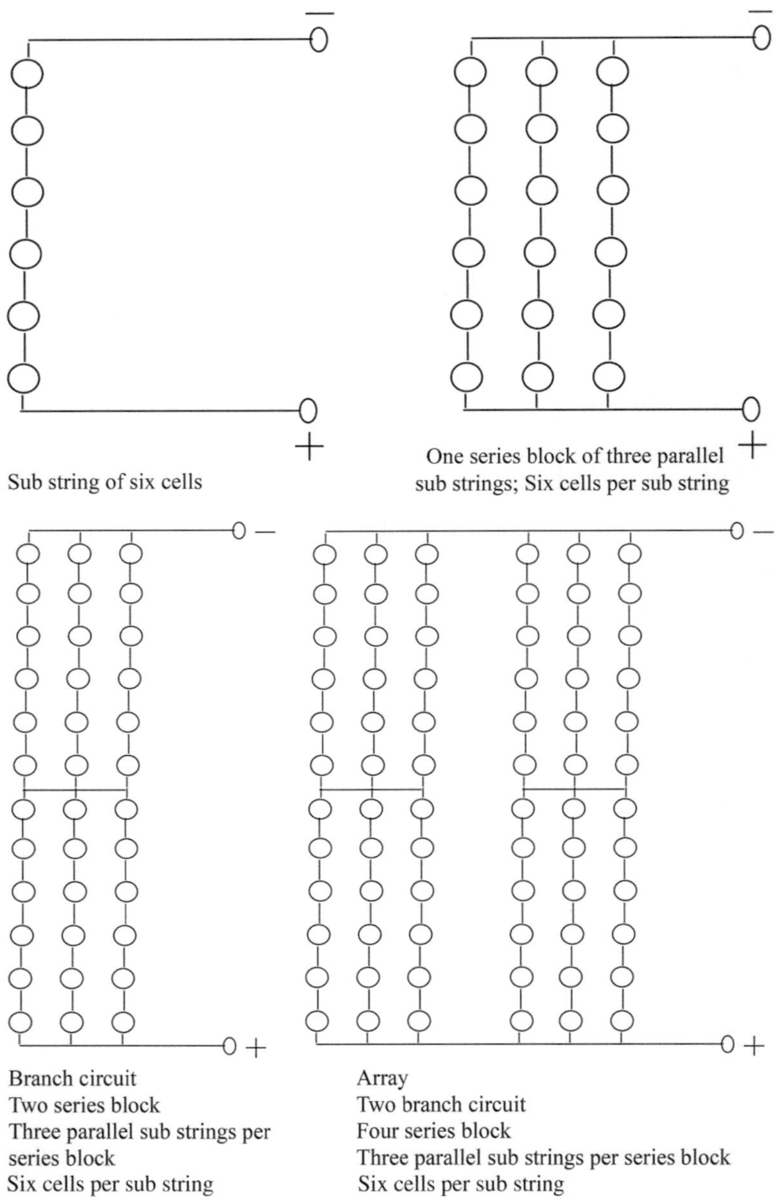

Fig. 7.1 Arrangement of solar cells in string, series block, branch circuit and array

modules, which are beginning to enter the market, require less material to produce than the thick crystalline products and so can be made less expensively. Today's commercial a-Si modules deliver 40–50 W/m^2 under full sun at STC. The modules are connected electrically in series and parallel to build arrays.

7.2 Mismatch Losses in Array Network

Solar cell fabrication procedure involve several controls; despite all automations the resultant cells possess significant manufacturer's tolerances. Performance-wise they exhibit variations in current-voltage characteristics. In particular, current (I_{max}) and voltage (V_{max}) of individual cells operating at maximum power (P_{max}) exhibit statistical distribution. So when mass-produced solar cells are connected in series and parallel within a network, the statistical distribution of cell parameter leads to electrical mismatch in the network; the mismatch limits the short circuit current of the combination to a value closer to the lower end of the distribution than to average. Consequently the maximum power output of the network is often less than the sum of the maximum power of individual cell; this is referred to as *mismatch loss*.

Similar results would occur with elements combined in parallel; open circuit voltage distribution will generate inter string power dissipation. This mismatch in fill-factor is not significant since the mismatch averages out producing no power loss.

In the above context a question of practical importance is: how much variations in cell characteristics can be allowed to meet the intended module and array performance specifications.

Several authors (e.g. Bucciarelli 1979; Gonzales and Weaver 1980; Chamberlin et al. 1995; Kaushika and Gautam 2001) have investigated the lowering of available power in solar PV arrays and found it to be due to small variations in cell characteristics. The mismatch losses in solar PV arrays can be due to a variety of reasons as follows (Sayed and Partin 1975; Treble 1991; Kaushika 1996).

- Manufacturer's tolerances in cell characteristics
- Environmental stresses
- Shadow problem

Physical differences between cells or cell processing material may arise during normal processing and different cells may have slightly different characteristic parameters. Consequently in an array the output power of the combination is less than the sum of maximum output power of constituents. This power loss may tend to enhance with time due to degradations resulting from aging.

The environmental stresses in solar PV array include the partial or full opening of a string due to cell cracking (provoked by hail impact) or otherwise. This not only gives rise to mismatch loss but also could result in excessive heating of the regions of power dissipation, which in turn can lead to solder melting and damage the encapsulant. The regions of excessive heating are referred to as hot spots (Sayed and Partin 1975; Feldman et al. 1981).

In the field, solar cells arrays are subject to shadows from both predictable sources as well as from such unpredictable sources as bird droppings or fallen leaves. The percentage power loss is much greater than the percentage of array area, which is shadowed; for smaller arrays with few or no parallel connections, one leaf could cause the system output to drop to a fraction of rated power, eventually resulting in system failure. Shadowed cells in series with illuminated cells block the current flow in entire series connection and tend to become reverse biased. Consequently, hot spots are developed in the localized regions of shadowed cell, which can damage the encapsulation or crack the cell causing module failure.

In this chapter, we investigate the effect of aging of solar cells in a module or array and hence reduction in available power. This investigation is based on real life data of ensemble of fresh and aging solar cells and is intended to enable the sizing of solar PV system in such a way that it can deliver power to the given load without failure for longer duration (Arab et al. 2004; Hussein et al. 2004; Kaushika et al. 2005).

7.3 Formulations of Fractional Power Loss in Solar PV Module and Array

7.3.1 Solar Cell Model

The solar PV module is formed by an appropriate series-parallel combination of solar cells that provides the required rated output voltage and current under normal conditions. Following Chapter 6 the electrical equivalent circuit of the solar cell is shown in Fig. 7.2.

Applying Kirchhoff current law, the solar cell terminal current, I can be expressed as

$$I = I_{ph} - I_D - I_{SH} \tag{7.1}$$

where I_{ph} is photo-generated current (Amp), I_D = diode current (Amp), and I_{SH} *is* shunt current (Amp).

The photo-generated current (I_{ph}) depends on both irradiance and temperature. It is measured at some reference conditions such as reference temperature, $T_{c,ref}$, reference radiation G_{ref} and reference photocurrent $I_{ph,ref}$ and is related as follows (Chenni et al. 2007; Rai et al. 2011):

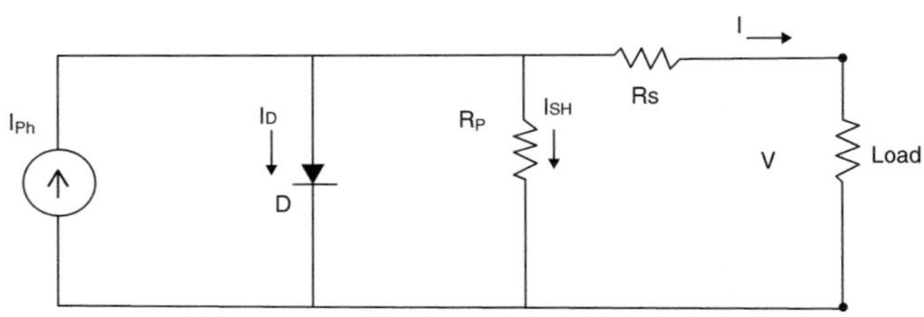

Fig. 7.2 Equivalent circuit of solar cell model

$$I_{ph} = \frac{G}{G_{ref}}\left(I_{ph,ref} - I_{SCC}\left(T_c - T_{c,ref}\right)\right)$$

where G is the actual solar radiation (W/m^2), T_c, the actual cell operating temperature of cell (K) and I_{scc} is the manufactured supplied temperature coefficient of the short circuit current (A/K).

The diode current is given by Shockley equation

$$I_D = I_0\left[\exp\left(\frac{e(V_c)}{\eta KT_c}\right) - 1\right] \tag{7.2}$$

where V_c is voltage across diode (V); I_0, reverse saturation current (A); η, diode ideality factor; R_s, series resistance (Ω); e, electron charge 1.602×10^{-19} C and K is Boltzmann constant, 1.38×10^{-23} (J/K) whereas the shunt current, I_{SH} is given by

$$I_{SH} = \frac{V_c}{R_p} \tag{7.3}$$

R_p is the shunt resistance (Ω).

The Eq. (7.1) can be written as

$$I = I_{ph} - I_0\left\{\exp\left[\frac{e(V + I\ R_s)}{\eta\ KT_c}\right] - 1\right\} - \frac{V + I\ R_s}{R_p} \tag{7.4}$$

Let $V' = V + I\,R_s$ and $I' = I + \frac{V'}{R_p}$. Equation (7.4) becomes as

$$I' = I_{ph} + I_0 - I_0\left\{\exp\left[\frac{eV'}{\eta KT_c}\left(\frac{\bar{V}_{max}}{\bar{V}_{max}}\right)\right]\right\} \tag{7.5}$$

Divide the above equation by $\bar{I}_{max}$

$$\frac{I'}{\bar{I}_{max}} = \frac{I_{ph} + I_0}{\bar{I}_{max}} - \frac{I_0}{\bar{I}_{max}}\left\{\exp\left[\frac{eV'}{\eta KT_c}\left(\frac{\bar{V}_{max}}{\bar{V}_{max}}\right)\right]\right\}$$

$$\text{Let}\,\alpha = \frac{I_{ph} + I_0}{\bar{I}_{max}},\quad \beta = \frac{I_0}{\bar{I}_{max}}\quad \text{and}\quad C = \frac{e\bar{V}_{max}}{AKT_c}$$

The above equation can be written as

$$\frac{I'}{\bar{I}_{max}} = \alpha - \beta\exp\left(C\frac{V'}{\bar{V}_{max}}\right) \tag{7.6}$$

where $\bar{V}_{max}$ is average voltage at maximum power, $\bar{I}_{max}$ = average current at maximum power and C is parameter related to fill-factor of solar cell.

This expression is similar to that of Bucciarelli's model (1979) with parameter α, β and *C*. It is assumed to provide an adequate expression of I-V curve near maximum power point. The parameter *C* can be matched with the curvature of I-V curve in the neighbourhood of maximum power point. It can also be expressed in terms of cell fill-factor. Thus in Bucciarelli's model series and parallel resistances end up encapsulated in the parameter *C*. Furthermore, by definition we know that the fill-factor (Treble 1991) is given by

$$\text{Fill factor} = \frac{P_{\max}}{I_{sc}\ V_{oc}} = \frac{V_{\max}\ \mathrm{I}_{\max}}{I_{sc}\ V_{oc}} \tag{7.7}$$

where I_{sc} = short circuit current of solar cell (Amps) and V_{oc} = open circuit voltage of the cell (Volts).

The maximum power point current $I_{\max}$ and maximum power point voltage $V_{\max}$ will satisfy Eq. (7.6) and can be written as follows:

$$\frac{\mathrm{I}_{max}}{\bar{I}_{max}} = \alpha - \beta \exp\left(C\frac{V_{max}}{\bar{V}_{max}}\right) \tag{7.8}$$

$$\text{For maximum power, } \frac{d(IV)}{dV} = 0$$

$$\text{or, } I + V\frac{dI}{dV} = 0$$

$$\text{and at maximum power, } I_{\max} + V_{\max}\frac{dI_{\max}}{dV_{\max}} = 0 \tag{7.9}$$

From Eqs. (7.8) and (7.9) we have

$$\beta = \frac{\frac{I_{\max}}{\bar{I}_{\max}}\left[\exp\left(-C\frac{V_{\max}}{\bar{V}_{\max}}\right)\right]}{C\frac{V_{\max}}{\bar{V}_{\max}}} \tag{7.10}$$

Substitution of the value of β in Eq. (7.9) gives

$$\alpha = \frac{\frac{\mathrm{I}_{\max}}{\bar{I}_{\max}}\left[\exp\left(C\frac{V_{\max}}{\bar{V}_{\max}} + 1\right)\right]}{C\frac{V_{\max}}{\bar{V}_{\max}}} \tag{7.11}$$

From Eqs (7.10) and (7.11) we see that in general I-V curve near the maximum power point may be represented by Eq. (7.6) wherein the α and β are non-dimensional quantities which may be defined in terms of I_{max} and V_{max}.

The fill-factor (*ff*) in terms of *C* can be written as (Bucciarelli 1979)

$$ff = \frac{P_{\max}}{I_{sc}\ V_{oc}} = \frac{V_{\max}\ \mathrm{I}_{\max}}{I_{sc}\ V_{oc}} = \frac{C^2}{(1+C)[C\ +\ \ln\ (1+C)]} \tag{7.12}$$

Table 7.1 Fill-factor and cell characteristics

S. No.	Fill-factor	C
1	0.3	1
2	0.4	1.72
3	0.5	2.85
4	0.58	4.22
5	0.67	6.83
6	0.7	8.13
7	0.75	11.8

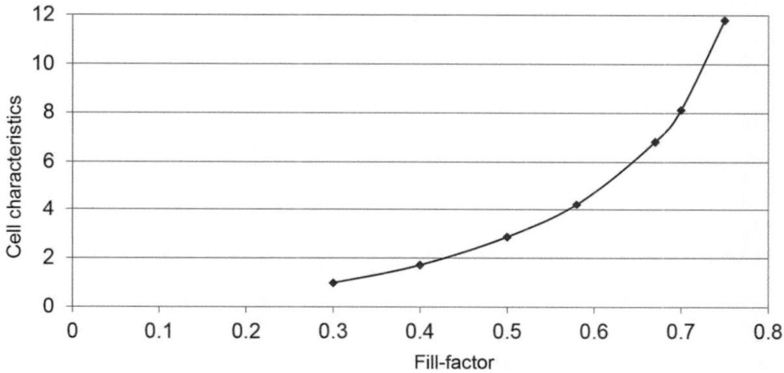

Fig. 7.3 Fill-factor vs. cell characteristics

The Eq. (7.12) is an implicit relation; it may be used to evaluate C as a function of *ff*. Some typical values are given in Table 7.1.

Figure 7.3 portrays the variation of cell characteristic parameter C with the fill factor; it shows that C lies in the range 8–11 for the ensemble of fresh cells.

7.3.2 *Fractional Power Loss in a Module*

The fractional power loss due to mismatch for L solar cells connected in series operating at common current is given by (Bucciarelli 1979)

$$E[\Delta P] = \frac{C+2}{2}\sigma_\eta{}^2\left(1 - \frac{1}{L}\right) \tag{7.13}$$

The fractional power loss due to mismatch for M solar cells connected in parallel operating at common voltage is given by (Bucciarelli 1979)

$$E[\Delta P] = \frac{C+2}{2}\sigma_{\xi}{}^{2}\left(1 - \frac{1}{M}\right) \tag{7.14}$$

Now consider a network of N_b series blocks of M parallel sub strings and each sub string having L cells in series as shown in Fig. 7.4. The fractional power loss due to mismatch is given by (Bucciarelli 1979)

$$E[\Delta P] = \frac{C+2}{2}\left[\sigma_{\eta}{}^{2}\left(1 - \frac{1}{T}\right) - \left(\sigma_{\eta}{}^{2} - \sigma_{\xi}{}^{2}\right)\frac{N_b}{T}(M-1)\right] \tag{7.15}$$

where $T = LMN_b$ i.e. total number of cells in the network.

From these formulations it may be noted that mismatch losses in both the series and parallel arrays are a sensitive function of the fill factor, variance in distribution of

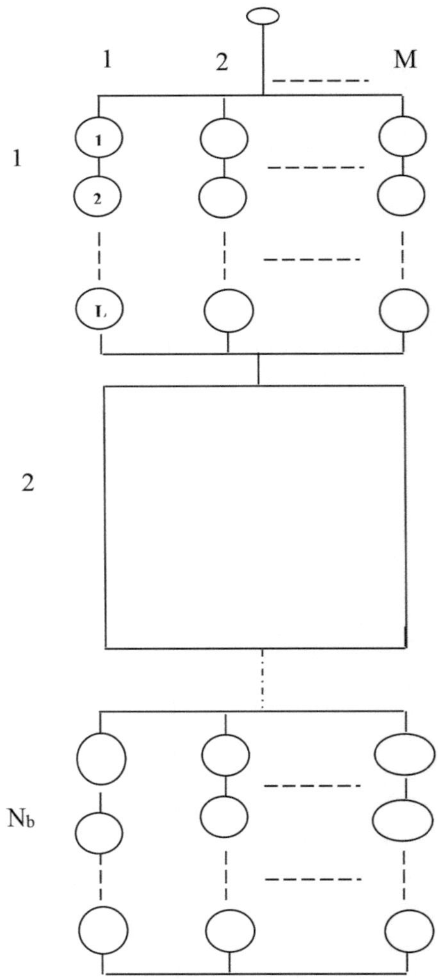

Fig. 7.4 Network of N_b series blocks of M parallel sub strings and each sub string having L cells in series (Kaushika and Rai 2007)

maximum power parameters of solar cell ensemble and total number of cells connected in the network. Gautam and Kaushika (2002a) have developed an efficient algorithm to simulate the performance of solar PV arrays.

7.4 Computational Results and Discussion

The formulations of mismatch losses are valid if the deviations, in cell max-power current and voltages, from the mean values for the ensemble of cells used in the network are small (limited to a few percent of their mean values). In the present investigation we have considered the mono-crystalline solar cells manufactured by Central Electronics Limited (CEL), New Delhi that has been manufacturing silicon solar cell for about three decades. Upon our request they supplied us the data for open circuit voltage (V_{oc}), short circuit current (I_{sc}), fill factor and maximum power parameters of constituent cells of two modules each made from an ensemble of 36 cells manufactured by them. CEL, New Delhi had also supplied to us yet another ensemble of cells that at the time of this investigation are representative of aging cells with impaired performance resulting from erosion at the top layer of the cell. These cells have been synonymously referred to aging or soiled cells in the present work. The variance and coefficient of variation (*CV*) of maximum power parameters of the ensembles are given in Table 7.2.

The standard deviation is found to be 2.5% of mean value for I_{max} for the fresh cells. The values of *CV* are derived from the following relationships:

$$\sigma_\eta = \frac{\sigma_I}{\bar{I}_{\max}} \tag{7.16}$$

Similarly σ_ξ of the distribution of maximum power voltage is given by

$$\sigma_\xi = \frac{\sigma_V}{\bar{V}_{\max}} \tag{7.17}$$

The fill factor, in general, varies from 0.5 to 0.75 for the mono-crystalline silicon solar cells. For example, for the cells manufactured by Central Electronics Limited (CEL) in India, it ranges from 0.70 to 0.75. The fill-factor was found to deteriorate to a value as low as 0.30 as a result of aging of cells.

It may be noted from Table 7.2 that the coefficient of variation (*CV*) is larger for I_{max} (fresh as well as aging cells) than the *CV* for V_{max} by a factor of 2.12 and 3.22 respectively for fresh and aging cells. Computational results of fractional power loss as a function of the number of cells and the variance of the distribution of maximum power parameters are illustrated in Figs. 7.5a and 7.5b). Both series and parallel

Table 7.2 Variance of cell max-power current and max-power voltage of fresh and aged cells

Category	σ_η^2	σ_ξ^2	σ_η	σ_ξ
Fresh cell	2.7×10^{-5}	6×10^{-6}	5.2×10^{-3}	2.45×10^{-3}
Aged cell	3.36×10^{-3}	3.24×10^{-4}	0.058	0.018

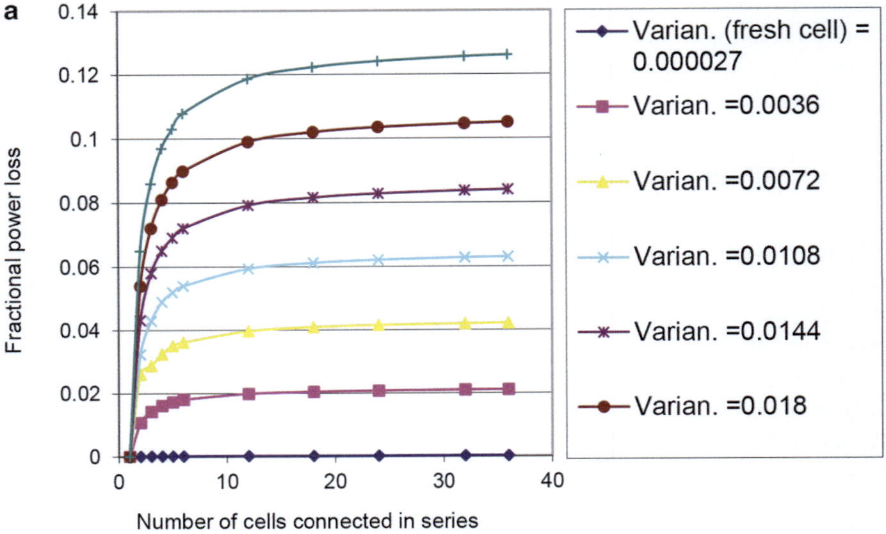

Fig. 7.5a Fractional power loss vs. number of cells connected in series for different variance of cell max-power current (Kaushika and Rai 2007)

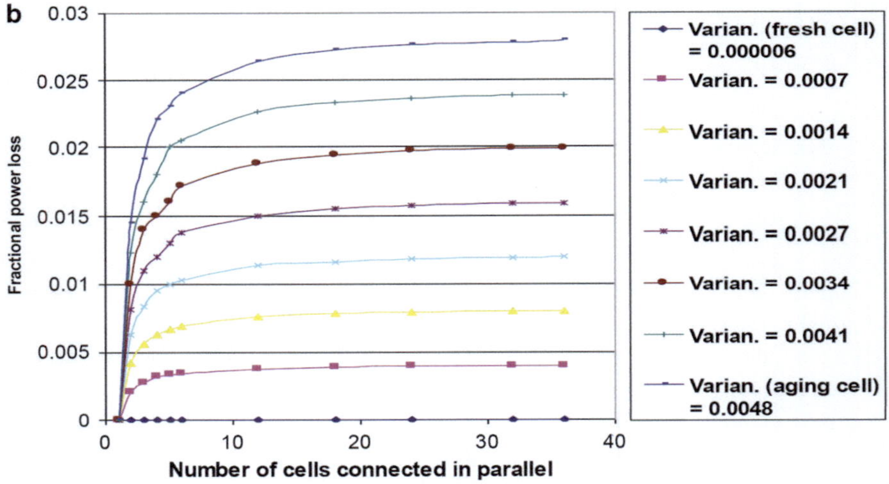

Fig. 7.5b Fractional power loss vs. number of cells connected in parallel for different variance of cell max-power voltage (Kaushika and Rai 2007)

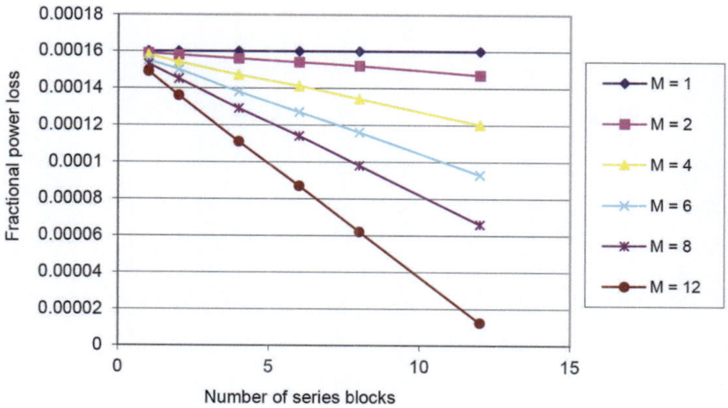

Fig. 7.6 Fractional power loss vs. number of series blocks for different parallel sub strings (Kaushika and Rai 2007)

interconnection schemes have been considered. Variance in $I_{max,}$ σ_η^2 is relevant in series connection and the variance in V_{max}, σ_ξ^2 is relevant in parallel combination; hence considered accordingly. The *CV* for I_{max} has greater impact on power loss in series string than *CV* for V_{max} has for parallel string. However, comparing Figs. 7.5a and 7.5b for the same values of *CV*, the fractional power loss is observed to be the same. For example in Fig. 7.5b for a *CV* of 0.0048 the fraction power loss is about 0.025–0.0275. From Fig. 7.5a, for *CV* of 0.0048 the corresponding fractional power losses lie in range of 0.02–0.04. The effect of series paralleling on the fractional power loss of network is investigated in Fig. 7.6. We have considered an ensemble of 144 ($T = 144$) fresh cells for the computation of fractional power loss for N_b series blocks of M parallel strings each having L cells given by $L = T/MN_b$. Results show that the fractional power loss tends to be lower for larger number of series blocks and higher values of M (number of parallel strings). For example, for $M = 1$, there is no variation with increase in N_b. Furthermore, for $N_b = 1$, the fractional power loss is due to paralleling and hence due to *CV* of V_{max} whose range of variation is very small. To illustrate the point let us consider a module made up of a series string of 36 cells. It will have $T = 36$, $N_b = 1$, $M = 1$ and $L = 36$; the fractional power loss (ΔP) is 0.00016. Now if it is series-paralleled with $N_b = 4$, $M = 3$, $L = 3$, the fractional power losses will reduce to 0.00011 (ΔP). In series string the power loss due to mismatch increases significantly with aging of solar cells. However, it can be reduced by an appropriate series paralleling arrangement. For example, for the ensemble of solar cells used in present work the fractional power loss of 12% resulting from the aging (soiling) can be counterbalanced by series paralleling in an array oversized by 2.4%. The variation of fractional power loss as a function of cell's lumped characteristic parameter, *C* is investigated in Fig. 7.7. It shows that aging cells are more sensitive to *C* and hence to fill-factor.

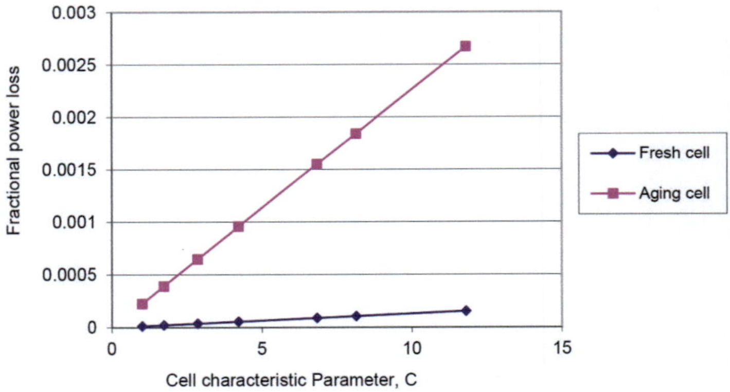

Fig. 7.7 Fractional power loss vs. cell characteristics (Kaushika and Rai 2007)

Chapter 8
BOS and Electronic Regulations

8.1 Balance of System (BOS)

The Balance of System (BOS) encompasses all components of solar PV system other than array. The array is the energy generating subsystem. It consists of PV modules, panel structures, interconnecting wires, junction boxes along with lightning arresters and associated accessories.

It generates variable DC only during sunshine hours which must be stored if the power is to be used during night or on a cloudy day. Solar generated electricity can be stored in battery bank. The battery banks can usually store sufficient electricity to power the load through two to five completely cloudy days and nights. Charge controller is yet another important component in connection with electrical storage subsystem like battery. These devices protect batteries from excessive charge when the module produces more electricity than the batteries can store. Also, they prevent deep discharge of batteries in no sun conditions. In absence of charge controllers, battery may suffer extreme wear and tear and become less effective, last for shorter time and possibly even short circuit. Sometimes during over charging the gassing from the acid of batteries may take place which in turn may become a fire hazard.

The output energy delivered from solar PV array to battery (load) may be maximized by using a maximum power point tracker (MPPT). The maximum power point tracker is used to correct the variations in the current voltage characteristics of array resulting from the variability of incident solar radiation and ambient temperature. It is basically a dc-to-dc converter that converts the input voltage of the load to a value coinciding with load line, without any change in the delivered power, but with a corresponding change in current.

The battery bank serves the load either directly or after conversion into AC. The battery power produced tends to be at a low voltage (12 V) and household application requires 240 V. An inverter is needed to convert direct current (dc) into alternating current (ac). Inverters are electronic devices employed in both stand-alone and utility-interactive PV systems to convert the direct current (dc) electricity

N. D. Kaushika et al., *Solar Photovoltaics*,
https://doi.org/10.1007/978-3-319-72404-1_8

produced by the PV array to alternating current (ac) common to the utility grid and required by most appliances and electronic devices. In grid connected systems, the dc power of the solar PV array is to be converted into AC power of fixed AC voltage. The AC should be sinusoidal of 50 Hz with no harmonics. Inverters are specified in terms of their capacity (in watts or kilowatts), their output voltage (e.g., 120 Vac) and their power quality. All utility-interactive PV systems require an inverter, as do stand-alone systems supplying ac loads. Stand-alone systems supplying only dc loads do not need inverters.

A meter is required to ensure that the system owner is credited for any excess power that is generated by the PV system and fed into the mains grid.

SPV power systems have been operated in various modes which can be classified into following three categories:

(a) Stand Alone: The entire power is generated by SPV array and stored in a battery to provide it in response to demand.
(b) Hybrid: In addition to SPV array such other means as wind generator and diesel generators are also used to produce power.
(c) Grid Connected Systems: In such systems the output of SPV plant is connected to the grid line and there is no storage battery.

The schematic diagram of a standalone PV power supply is shown in Fig. 8.1. In this system the solar array needs the use of an optimum load into which it can deliver its maximum power. But the requirement varies with insolation and temperature. The battery in turn offers a load to the array that varies with its state of charge and temperature. The mismatches of different system components offer problem in optimum sizing of the system and results in loss of energy and degradation of components. A matched interconnection of these power system components under variable conditions may be realized with the help of Maximum Power Point Tracker (MPPT). So in practice solar photovoltaic system requires several components in addition to PV modules. These are called Balance of System (BOS).

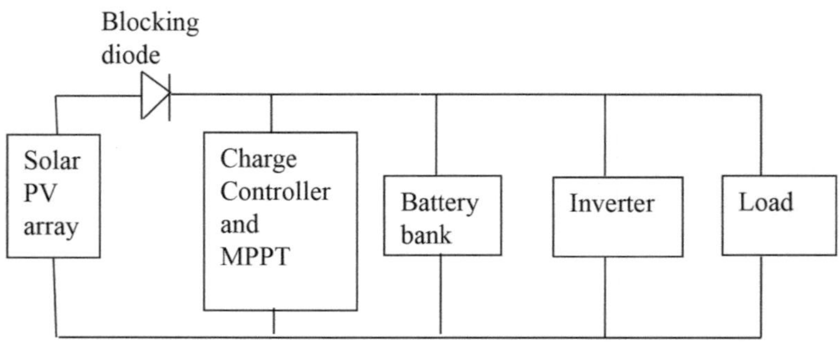

Fig. 8.1 Schematic of standalone photovoltaic power supply system

8.2 Classification of BOS

In view of the above, BOS includes dc-dc converter, inverter, charge controller, maximum power point tracker, solar irradiance sensor, electronics for energy management, electrical switches, circuit breakers, fuses, wires & cables, protection devices etc. depending on the technical specifications and requirements of the PV system. All the components of the balance of systems may be classified in three categories: Mechanical, Electrical and Electronics BOS.

8.2.1 Mechanical BOS

Mechanical BOS includes PV panel structures, battery racks, poles and stays for carrying electrical wires and cables etc. to withstand high wind speed and hail storms for stability.

8.2.2 Electrical BOS

Since a PV system is an electrical energy conversion unit, the electrical BOS subsystems are similar to those used in conventional electrical power plants. This includes lightning arresters in the field to control panels consisting of required cable and wiring, circuit breakers, switchgears and so on.

Batteries are needed in stand-alone systems to store electricity for use at night and during periods of low sun. Liquid-electrolyte, lead-acid storage batteries are, by far, the most widely used. Deep-cycle batteries are required for long-term service. Batteries are connected in series to provide the necessary system voltage, and in parallel to provide the required amount of storage rated in amp-hours at the system's voltage.

Captive-electrolyte, or "gel-cell", batteries are available for applications where liquid electrolyte is undesirable, but their initial cost is typically greater than for those with liquid electrolyte. Many new battery technologies are being investigated for use in electric vehicles. When commercialized, these new batteries will benefit PV systems as well.

String combiners are prefabricated junction boxes that are useful to facilitate the parallel connection of the series strings of PV modules, called source circuits, which make up the PV array. String combiners often have integral *lightning protection* devices, such as silicon-oxide varistors, which shunt damaging high-voltage transients to ground. Article 690 of the National Electrical Code (NEC), which governs

PV systems, requires over current protection and a disconnect means for PV array source circuits, and most string combiners include *over current protection* in the form of pull-out fuses or dc-rated circuit breakers, which also can serve as string disconnects for array testing and service.

Disconnect switches are needed for system servicing and personnel safety, as set forth in NEC Article 690. They are typically installed on the inverter's input and output, at the array output and on the battery bank's output. Most disconnect switches also include *over current protection*, either as fuses or circuit breakers. A fuse or other means of "catastrophic" over current protection is also necessary at the battery bank and, large battery banks even have fuses internal to their string wiring to isolate portions of the bank in the event of a fault. Be certain to employ dc-rated components in all dc circuits.

There are some main components that are key to any PV system, assuming the PV array is mounted and the necessary cables and switches are in place. This equipment is often referred to as balance of system equipment required to complete the system. This includes some of the equipment explained below, and wires, conduit, a grounding circuit, fuses, safety disconnects, outlets, metal structures for supporting the module and any additional materials that are part of the PV system.

Batteries are sometimes necessary to store electricity for use during times when it is needed. Or to help meet loads when PV modules are not generating sufficient power to meet load requirements. Electricity storage is a critical component of many PV systems. If the power is needed at night or on a cloudy day, solar generated electricity can be stored in batteries. These battery banks usually store sufficient electricity to power the home through two to five completely cloudy days and nights.

8.2.3 Electronic BOS

The electronic BOS is required to monitor, convert, manage and control the PV generated energy to the loads. Commonly used electronic BOS in solar PV system may be divided into following categories.

8.2.3.1 Energy Conversion Subsystem

Energy conversion subsystem may have dc/dc converters and MPPTs; or dc/ac inverters.

Dc/Dc Converters and MPPTs

Since the output of a solar PV array is variable dc, a dc-to-dc converter is required to convert the dc output from the array to a voltage suited to the requirement of the battery or load. A dc-to-dc converter may be located between the solar PV array and battery bank to ensure optimum power. Among the dc-to-dc converters, the boost converter is preferred because it can draw current continuously at the input side. The voltage for optimum power transfer varies with insolation and array operating temperature. A feedback control is required in order to ensure that the power converter is operating at optimum voltage transformation ratio.

A maximum power point tracker (MPPT) is required to minimize the mismatch between the solar PV array output and load. In stand-alone solar PV system without battery storage, the module or array generates variable dc output during the day and delivers it to the dc load. In such a simple system, the amount of energy delivered to the load depends not only on the incident solar radiation and ambient temperature but also on the operating load on the array current-voltage characteristics. If the load is so chosen that the array operating point is at its peak power point during the mid-day the array then delivers maximum possible amount of energy at that time. But since both the incident solar radiation and the ambient temperature vary during the course of a day, the position of the array peak power point varies with time.

Therefore, it is quite unlikely that the operating point will be coinciding with the array peak power point at all times. There is some loss in the delivered energy due to mismatch between the locus of load line and locus of peak power points. The power loss due to load line mismatch can be overcome by using a maximum power point tracker in the system.

A maximum power point tracker is an essentially dc-dc converter, which continuously monitors the output current and voltage from the module or array as long as there is enough sunshine and module is generating electricity. At the same time it computes the peak power available, and then it converts the input dc voltage to the load to a value coinciding with the load line without any change in the delivered power, but with a corresponding change in the current Thus, the peak power at any instant is fed to the load at a voltage appropriate to coincide with its load line. The possible power loss due to mismatch is partially eliminated.

It has built-in control logic, usually operated by a microprocessor that senses the array voltage and current at frequent interval (typically every 30 ms), computes the power output and compares it with the previous value. If the power output has increased, the array voltage is stepped in the same direction as the last step.

If the power has decreased, the array voltage is stepped in the opposite direction. Eventually, the array voltage reaches near the maximum power point and remains constant irrespective of irradiance, temperature and load. The power consumption of MPPT is typically 3–7% of the array output. Many techniques for maximum power point tracking of SPV arrays are now known which include:

(i) Hill climbing/Perturb and Observe method
(ii) Incremental Conductance method
(iii) Neural network
(iv) Fuzzy Logic Control
(v) Fractional Open Circuit Voltage method
(vi) Fractional Short Circuit Current method
(vii) Ripple Correlation Control (RCC) method

(viii) Current Sweep method
(ix) DC-Link Capacitor Droop Control method
(x) Load Current or Load voltage Maximization method
(xi) *dP*/*dV* or *dP*/*dI* Feedback control method
(xii) State-based MPPT method
(xiii) One Cycle Control method
(xiv) Slide Control method

The Perturb and Observe method as well as the Incremental Conductance algorithms are the most applied techniques. Esram and Chapman (2007) have reviewed all the above mentioned techniques, which can be broadly classified in two categories: PV array dependent MPPT techniques and PV array independent MPPT techniques.

A summary of PV array dependent and independent MPPT techniques is shown in Tables 8.1a and 8.1b respectively.

Table 8.1a Characteristics of PV array dependent MPPT techniques

MPPT technique	True MPPT?	Analog or Digital?	Periodic tuning?	Convergence speed	Sensed parameter
Neural network	Yes	Digital	Yes	Fast	Varies
Fuzzy Logic control	Yes	Digital	Yes	Fast	Varies
Fractional V_{OC}	No	Both	Yes	Medium	Voltage
Fractional I_{SC}	No	Both	Yes	Medium	Current
Current Sweep	Yes	Digital	Yes	Slow	Voltage, Current
State-based MPPT	Yes	Both	Yes	Fast	Voltage, Current
One Cycle control	No	Both	Yes	Fast	Current

Table 8.1b Characteristics of PV array independent MPPT techniques

MPPT technique	True MPPT?	Analog or Digital?	Periodic tuning?	Convergence speed	Sensed parameter
Hill-climbing/P & O	Yes	Both	No	Varies	Voltage, Current
Incremental conductance	Yes	Digital	No	Varies	Voltage, Current
RCC	Yes	Analog	No	Fast	Voltage, Current
DC Link Capacitor Droop Control	No	Both	No	Medium	Voltage
Load Current or Load Voltage maximization	No	Analog	No	Fast	Voltage, Current
dP/*dV* or *dP*/*dI* Feedback control	Yes	Digital	No	Fast	Voltage, Current
Slide control	Yes	Digital	No	Fast	Voltage, Current

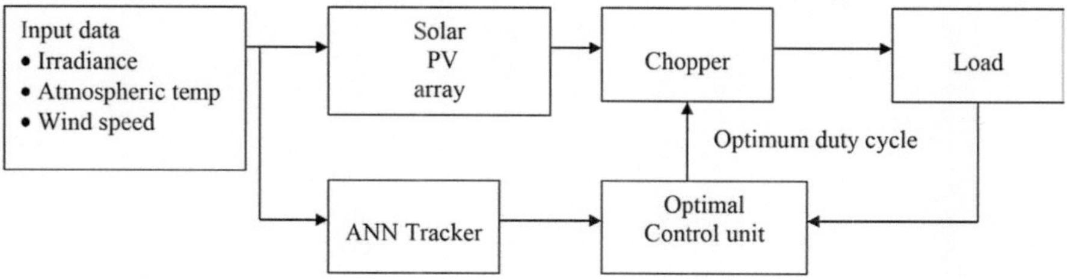

Fig. 8.2 Schematic diagram of ANN controlled SPV system

Rai et al. (2011) have discussed an artificial neural network (ANN) based maximum power point tracking controller. The schematic block diagram of such a controller is shown in Fig. 8.2. The controller consists of ANN tracker, optimal control unit and dc-to-dc converter. The ANN tracker estimates the voltages and currents corresponding to maximum power delivered by solar PV array for variable cell temperature and solar radiation. The cell temperature is considered as a function of ambient air temperature, wind speed and solar radiation. The ANN model consists of three layers consisting one input layer of two neurons, one hidden layer of nine neurons and one output layer of two neurons. The tracker is trained employing a set of 124 patterns using back propagation algorithm in MATLAB. The mean square error of tracker output and target values is set to be of the order of 10^{-5} and the successful convergent of learning process takes 1281 epochs (Fig. 8.3). The accuracy of ANN tracker has been validated employing different test data sets. The output of the control unit is a combination of voltage and current estimated by ANN tracker and load resistance.

An ANN model of the maximum power point tracking controller is simulated with SIMULINK (Fig. 8.4). The temperature and radiation block shows the input layer. The individual inputs are weighted by corresponding elements of the weight matrix (input_weights). The resultant product is added with bias matrix (input_bias) that is transformed by hyperbolic tangent sigmoid transfer function (tansig), which is the input to the hidden layer. The neurons of hidden layer are applied to the output layer, which is approximated by a linear transfer function (purelin). Figure 8.5 shows the actual data and the ANN-estimated maximum power current and voltage variations with the temperature and radiation. One can notice the good agreement between the two sets of results. The relative error between experiment data and simulated values on maximum power point is less than 5%.

Dc/Ac Inverters

A dc-to-ac inversion is required for ac loads. Inverters are broadly classified into two categories: stand-alone inverter and grid tied inverter. In stand-alone system, the inverter controls the voltage and frequency of the AC; sometimes called off grid inverter. The grid tied inverters are intended to feed power into AC supply. The

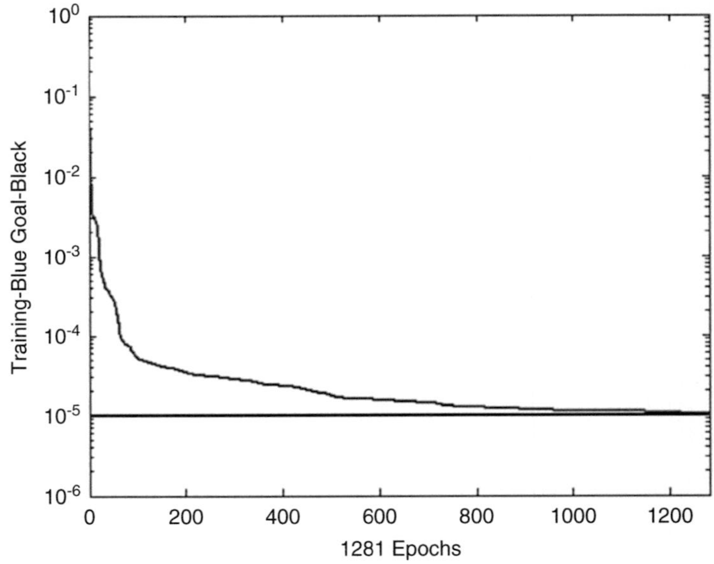

Fig. 8.3 Mean square error vs. number of iteration

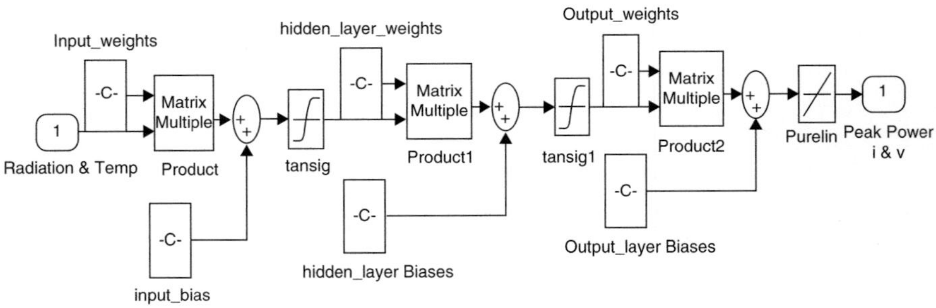

Fig. 8.4 ANN model of the maximum power point tracker implemented with Simulink

difference between the two categories is because of control algorithm which is implemented using microprocessor/microcontroller.

An inverter converts the dc from the array or battery bank to single-phase or three-phase ac as per load requirement. Quality inverters provide sine-wave electrical output that conforms to utility power quality standards in terms of harmonic content and voltage and frequency stability established for the power grid. All utility-interactive inverters must meet minimum power quality standards set by the American National Standards Institute (ANSI).

In modern solid-state inverters two techniques are mainly used to get a sinusoidal output:

(i) Pulse-width modulation (PWM): In this technique, sinusoidal waveform is generated by switching square-wave inverter stages at a rate higher than the

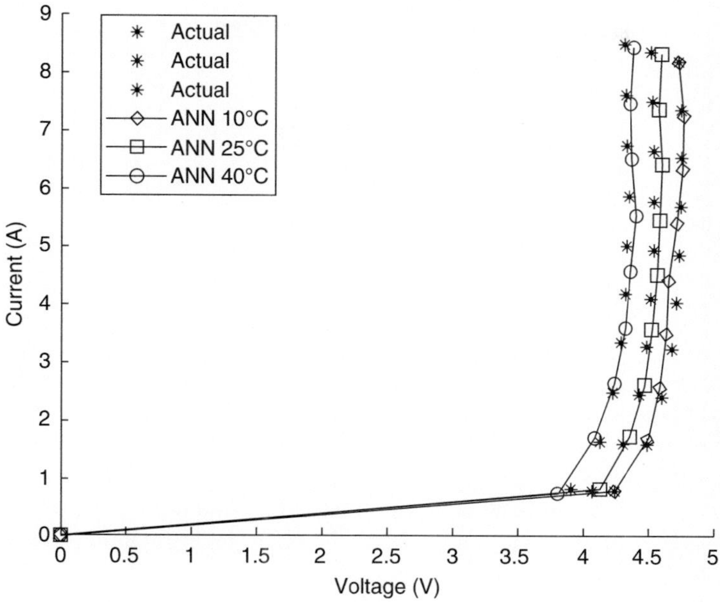

Fig. 8.5 Maximum power current vs. maximum power voltage for different values of temperature and radiation

fundamental frequency. The variation in the conduction time of the power switches (pulse width) controls the output voltage. The total harmonic distortion (THD) in the inverter output is inversely proportional to the switching rate.

(ii) Waveform synthesis: In this technique, a stepped output waveform approximating to a sine wave is constructed by combining the phased outputs of several inverter stages at the fundamental frequency switching. Higher level inverters are used to lower the harmonic distortion and better approximation to sine wave.

8.2.3.2 Energy Control and Management Subsystem

This subsystem may have power condition and synchronizer, charge controller to protect the battery bank from overcharged or deep discharge, timer, load switching, monitoring, data logging etc. for control and management.

The charge controller regulates the flow of energy between the solar PV array and the battery bank. These electronic devices prevent damage to the batteries by protecting against both overcharging by the PV array and deep discharging by the load. In the absence of charge controllers, the battery may suffer extreme wear and tear and become inefficient. The life of the battery will be short and in worst condition a short circuit may occur. In the case of standalone solar PV system with battery storage, the floating condition occurs when battery voltage fluctuates

Table 8.2 Summary of regulation process

S. No.	Constraints	Command	Remarks
1	If $V_b > V_{max}$	Disconnect battery from PV array	
2	If $V_b > V_{max}$ and $V_b < V_{t1}$	Reconnect the battery to the PV array	$V_{max} - V_{t1}$ gives the hysteresis voltage
3	If $V_b < V_{min}$	Disconnect the battery from load	
4	If $V_b < V_{min}$ and $V_b > V_{t2}$	Reconnect the battery to load	$V_{min} - V_{t2}$ gives hysteresis voltage

between maximum (100%) and minimum (50–60%) state of charge (SOC) of the battery. This type of system is auto regulated during the day and at night the regulation makes use of blocking diode which blocks the reversal of power from battery to PV array.

In the case of overcharge and deep discharge conditions the battery voltage reaches at some critical values. In that case the electronic regulation is characterized by two functions:

(i) Knowing the constraints (V_{min} and V_{max}).
(ii) Collecting information on the state of charge of the battery (battery voltage, V_b).

The summary of the regulation process are given in Table 8.2. A number of different models are available, from basic to sophisticated one. Desirable features include low-voltage load disconnect, field adjustability and provisions for a finishing charge. Because battery charging characteristics change with temperature, battery-temperature compensation is recommended when selecting a charge regulator in all but the most-basic systems.

8.3 Electronic Regulators

The solar PV array has an optimum load into which it can deliver its maximum power which varies with solar insolation and temperature. The battery offers a load to the array by providing a bias to it that varies with battery state of charge and temperature. The load derives power from the battery and changes its state of charge which varies with load pattern. The interconnection of these power system components must be designed to have them matched under all conditions. For small or intermediate applications, stand-alone PV systems are self-sufficient in generating, storing and supplying electrical power to local electrical loads.

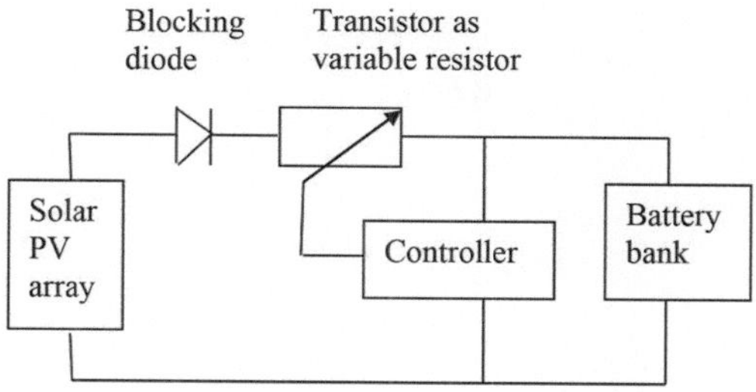

Fig. 8.6a Series regulator

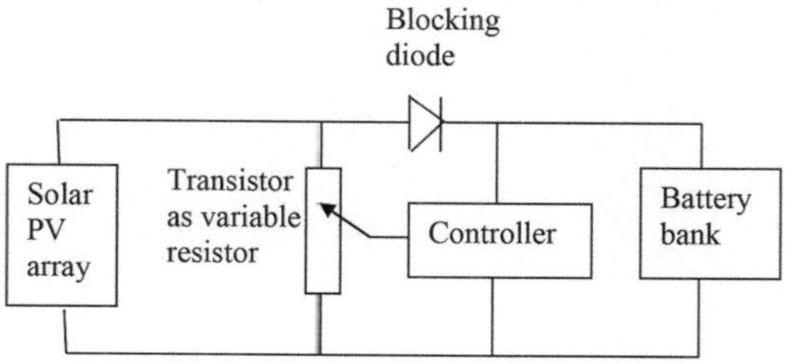

Fig. 8.6b Shunt regulator

The electronic regulator may be either a stand-alone device or a control circuitry integrated within a battery-powered device, or battery recharger. In PV array, basically there are two methods by which the voltage and current can be controlled either using series regulator or shunt regulator. In series regulator, a resistance is introduced in series with array to reduce the load current and allowing the array voltage to rise up to its open circuit value whereas a shunt regulator dumps currents from pulling down both the load and array voltage (Figs. 8.6a and 8.6b). In both the regulators a blocking diode is required to prevent the battery bank being discharged either through the array or through the shunt regulator. These regulators may be implemented by using transistors as variable resistor whose value is a function of the state of charge of the battery. This approach requires the transistor to dissipate a large amount of power and is generally not used.

The regulation can also be achieved if the transistor is used as switch which turns on and off in such a way that the mean current flowing to the battery is reduced (Figs. 8.7a and 8.7b). In the series switch regulator, the switch connects and disconnects the current to battery bank. The mean current depends on the duration for which the switch is on. If the switching is fast the voltage across the battery will be constant. The excess power from the PV array is dissipated in cells of the PV array.

The shunt switching regulator also operates in similar manner. The blocking diode prevents the battery bank being discharged through the switch and also acts as a series switch which is off when the controlled switch is on. The excess power is dissipated in the cells of the PV array.

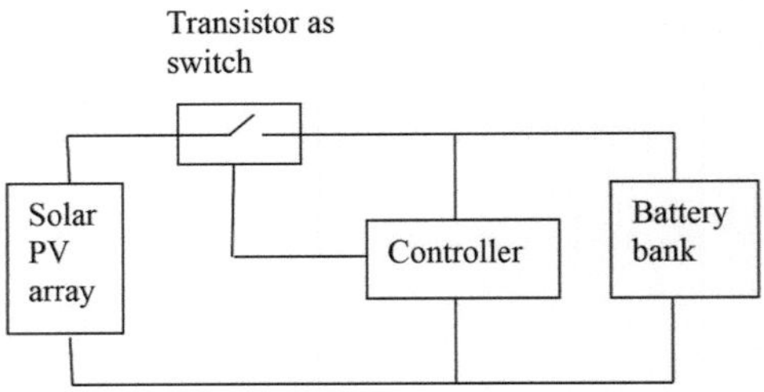

Fig. 8.7a Series switch regulator.

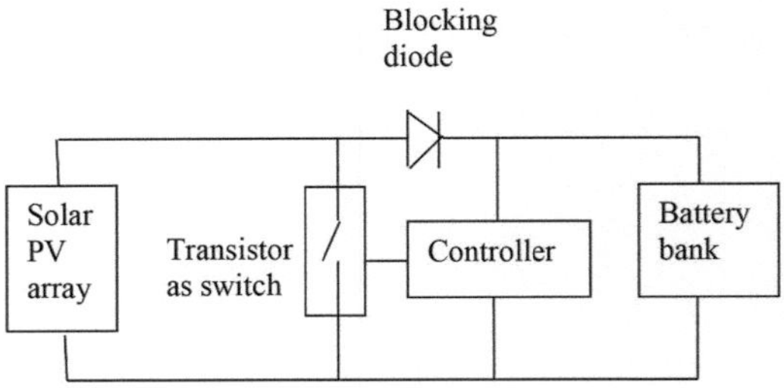

Fig. 8.7b Shunt switch regulator.

Chapter 9
Repertoires of Applications

9.1 Space Applications

The solar conversion efficiency of first silicon solar cell was reported to be 6%. The estimated cost of production of solar cell was USD 600 per Wp. The applications that could be envisioned at that time were of power generation in remote locations where fuel could not be easily delivered. The obvious application was satellites where the requirement of reliability and light weight made the cost of solar PV unimportant. The first ever application of solar photovoltaics in space was in 1958 in Vanguard-I satellite. A small solar PV power unit of 6 mW, developed by space engineers at United States Army Signal Corps (USASC), was used as a power source in a back up radio transmitter. It was a rather successful trial and it sped the laboratory test and evaluation research in area of solar PV. Simultaneously (1958 onwards) in USSR solar PV units were used to power the entire electronics of satellites. These units operated successfully for 2 years. I960 onwards American satellites also started using solar PV units to power the electronic systems. The maximum capacity was of 150 Wp which paved the way for further capacity increase of solar PV power units in satellites as follows:

Name of mission	Year	Solar PV power unit (Wp)
Nimbus	1964	500
Orbiting astronomical observatory	1966	1000

Extensive space applications were followed by the growth of market for space programme and research inputs which in turn resulted in solar PV power cost reduction to about USD 100–200 Wp. This price also was high to support mass scale terrestrial applications.

N. D. Kaushika et al., *Solar Photovoltaics*,
https://doi.org/10.1007/978-3-319-72404-1_9

9.2 Terrestrial Applications

Large scale solar PV application was triggered in early 1970 due to rapid escalation of the cost of fossil fuels and increasing environmental concerns about their use.

In the initial stages solar PV was used for such assorted applications as solar lantern in hilly areas and street lights in remote villages. In 1980 some commercial companies came in the field of solar PV application such as in India Tata-BP Solar. They identified several need-based clients such as:

1. Illuminating hamlets in scorching deserts;
2. Hospitals in far-flung areas using solar refrigeration for polio drops preservation;
3. Agricultural water pumping in villages which were not connected with grid; and
4. Distant towns entrepreneurs using solar PV power for furtherance of their business opportunities (Figs. 9.1a and 9.1b)

Fig. 9.1a Solar PV power for illuminating hamlets in scorching deserts (Adapted from Tata BP Solar promotion pamphlet)

Fig. 9.1b Solar PV power for remote refrigeration (Adapted from Tata BP Solar promotion pamphlet)

This awareness of PV technology is rising and there are an increasing number of systems beginning to appear as the concern for the need of 'clean' electricity is gaining importance. The global SPV module in 1989 was 45 MW. It has now increased to over 200 GWp. Some typical applications for photovoltaics today are as follows:

Lighting and security and other consumer products
Village power
Remote refrigeration
Electricity for individual buildings or entire communities
Water pumping and irrigation
Supplementing the utility grid for distribution support
Communication
Remote monitoring equipment
Aids to navigation
Cathodic protection
PV/Diesel hybrid generators
Military applications

9.3 Cost Effective Applications

The reduction in cost of solar PV power has been the aim of some applications during the last decades. These applications are described hereunder.

9.3.1 Solar Photovoltaics for Buildings

The distributed PV systems on buildings may be considered as an attractive application to attain commercialization of solar PV power on the utility grid. It could be a preferred power source for buildings in remote areas not served by the utility grid. The PV systems on buildings would also avoid the cost of land required for ground-mounted systems, as well as the cost of site development, foundations, structural support systems, underground electrical distribution and the utility connection. In practice the PV modules would become an integral part of the building, often serving as the exterior weathering skin. The building would provide the PV aperture area and support structure, and the building's utility service would become the consumer of PV electricity. In addition, PV systems on buildings would become a reliable power source in areas plagued with frequent power cuts. The design and installation of effective and efficient solar PV depends on several parameters as follows:

(i) *Location:* The amount of sunlight available will affect the amount of electricity generated.

(ii) *Orientation:* The orientation of a building and positioning of solar arrays are important factors to maximize energy production. In the northern hemisphere, a south facing façade will collect maximum solar energy throughout the year and produce the maximum electricity.
(iii) *Tilt angle:* The solar incident angle (θ) on a surface varies with declination (time of the day and day of the year) and latitude of the location. At noon time (time of maximum irradiance) it is expressed as

$$\theta = \phi - \delta$$

where ϕ = latitude of the site and δ = declination angle.

So, several options are possible for the mounting of modules/array. It may be placed at fixed tilt angle to maximize the generation of yearly average energy. The tilt may be seasonally adjusted in accordance with the energy requirement in summer, equinoxes and winter respectively. Owing to cost considerations, fixed tilt arrays are often used.

The above parameters can easily be optimized in the design of a building. The inclined building roof or the appropriate building wall may be chosen to maximize the electricity production. The resultant systems are known as Building-integrated PV systems (BIPV). The BIPV systems, therefore, have the potential to be architecturally attractive, which adds to the commercial acceptance of the technology on the part of design professionals, building owners, and the general public.

Furthermore, BIPV components replace conventional building materials and labour, thus further reducing the installed cost of the PV system.

With reduced installation costs, improved aesthetics and all the benefits of distributed generation, building-integrated PV systems have become an attractive option around the world for both residential and commercial buildings.

The BIPV power systems have been operated in various modes which can be classified into following three categories:

(a) *Stand-alone:* The entire power is generated by SPV array and stored in a battery to provide it in response to demand.
(b) *Hybrid:* In addition to SPV array such other means as wind generator and diesel generators are also used to produce power.
(c) Grid-connected systems: In such systems the output of SPV plant is connected to the grid line and there is no storage battery.

Photovoltaic systems for buildings can be either stand-alone, hybrid or grid-connected. In a stand-alone system, the building has no connection to the utility grid and often relies on a bank of batteries to store power for use at night and during periods of limited sun. In India the hybrid systems have been considered to avoid excessive cost of solar panels and the battery bank. For example, a 10 kWp solar PV and Biomass Gasifier Hybrid Power Plant have been set up by Tata Energy Research Institute to serve the hostel and 10 numbers (PL-11) standalone dusk-to-dawn street lights. In a grid-connected or utility-interactive (UI) system, the building receives electricity from both the PV array and the utility grid. Some PV systems are capable

Fig. 9.2 Indira Parayavaran Bhawan, New Delhi

of both stand-alone, utility-interactive modes of operation and or hybrid mode. In India an innovative application of building integrated PV system has been demonstrated in office building of Ministry of Environment, Forest and Climate Changes. The building is referred to as Indira Parayavaran Bhawan (Fig. 9.2) and it is located at Jorbagh Road, Aliganj, New Delhi (28.58°N). This is net zero energy multi-storeyed building with 100% on-site power generation.

The details of the BIPV system are as follows:

Capacity of power generation:	930 kWp
Annual energy requirement:	14 lakh units (kWh)
Net energy consumption:	Zero
Total area of solar panels:	460 m^2
Photovoltaic panel:	Mono-crystalline (20% efficiency)
Nature of power generation:	Grid-Interactive

Owing to the advantages and promise of continued cost reductions, the potential for PV-powered buildings has escalated. In a best-case scenario, Arthur D. Little, Inc. (1995), has projected that the annual U.S. market for PV on buildings could exceed $2.5 billion in the next 10 years. Japanese and European markets, already under development, hold similar potential.

9.3.1.1 Photovoltaics as Glazing Material

Mounting PV panels on buildings have been carried out since the solar photovoltaics panels became commercially available. In the initial stages building clading with solar photovoltaic panels was carried out in retrofit manner. The traditional PV modules were manufactured with aluminium frames. These modules were designed for grouping on racks to be placed on buildings or on the ground for the reduction of building electrical load. The installation was often carried out in the post construction phase of the building. Installing the modules meant drilling holes in a south facing roof and installing them on expensive racks. In the event of roof repair the PV installation had to be removed. In recent years the demand for solar PV module as a building element has been recognized; the modules could then mimic windows or skylights or they can become part of the actual roofing material. This application has been referred to as building integrated photovoltaics (BIPV). Hundreds of projects in Europe, USA and Asia have been completed and several others are in the pipeline. In what follows we present case studies of some of these projects.

BIPV means that the solar features become part of the weather-proof skin of a building. The array takes over the normal function of roof and is not just placed on top of an existing structure. It serves both as a roof covering and as an electrical production device. Based on this concept, several BIPV R&D demonstration units have come up around the world. The photovoltaics websites give an introduction to the construction and performance of these BIPV roofs and walls. Some of these units that we have studied through website may be listed as follows:

(i) **BIPV at the School of Construction Management of University of Reading, UK**

The University of Reading was refurbished a few years ago. In this process south elevation of the roof was considered an excellent site for BIPV. It is unshaded and captures maximum solar energy in winter. The roof generates electric power directly and provides part of the electric supply to the building. The solar photovoltaic roof is also being used as an R&D demonstration facility for advance training and research in the school of construction and management and engineering.

(ii) **Folk Centre has conducted several projects in the field of building integrated photovoltaics**

The centre has good theoretical and practical know-how in the area. It offers general consultancy regarding production and marketing of building integrated photovoltaics, including installation, operation & maintenance and supervision of local production. The characteristics of some of these projects are as follows.

(A) **The Building integrated photovoltaics in "Skibsted Fjord" training centre**

The solar cells are integrated in the window panes applying the principles of security window panes. The PV cells, in the installation in the facades of Skibstedfjord, are placed in various patterns, and demonstrate the possibilities to use the cells as an architectural effect. The patterns can be used for shading as well as for decoration purposes.

(B) **Folk Centre main building**

Various organs of the main building have been integrated with solar photovoltaic panel wherein the panels have been architecturally integrated to serve some blind filters of sunlight in summer and generate electricity for use in the building.

Some of these projects have been funded by Danish energy authority. The intended objective of these projects is to demonstrate the use of photovoltaics as roofing which adds a new architectural dimension to solar cell. Until now solar photovoltaics have been retrofitted on top of existing structures.

(iii) **BIPV Projects in India**

In India also several BIPV projects have been commissioned. The notable of these projects are:

(A) 10 kWp solar PV Gasifier Hybrid Power Plant for Tata Energy Research Institute, Delhi hostel and 10 numbers (PL-11) standalone dusk-to-dawn street lights.
(B) Building integrated solar PV system plant at Mamata Machinery, Hyderabad.
(C) 36.3 kWp, hybrid PV power plant Matrimandir, Auroville, Pondicherry.

9.3.1.2 Solar PV Metal Roofing

Solar PV metal roofing means that solar features become part of weather-proof roof or wall of the building. Presently research and development activities in the field of building integrated photovoltaic systems are concentrating on the development of constituent products. It includes three general areas: integral roof modules, roofing tiles and shingles, and integral modules for vertical facades and sloped glazing. Many BIPV systems have been integrated into roofs because these have the highest solar exposure and so provide the highest power output. Recent innovations have also been made in regard to their use as vertical curtain walls and awnings because integrated PV modules can replace facade materials, such as granite and architectural glass. New semi-transparent thin-film modules are also being developed, so PV can even function as a skylight or a window. In solar PV metal roofing the factors that are often considered are as follows: The nature of PV technology, type of PV module, type of battery technology and characteristics of charge controller and inverter.

In India, BIPV systems have used three types of modules: solar PV modules, solar PV metal roofing modules and dummy modules. The salient features of dummy modules are without solar cells. Instead of cells the back tedler is screen printed with solar cell pictures which look similar to actual solar cells. The other related materials used shall remain same except solar cells. Whatever the method of integration, BIPV systems are often grid-connected.

This means that the solar power is fed into the existing mains electricity supply of the building.

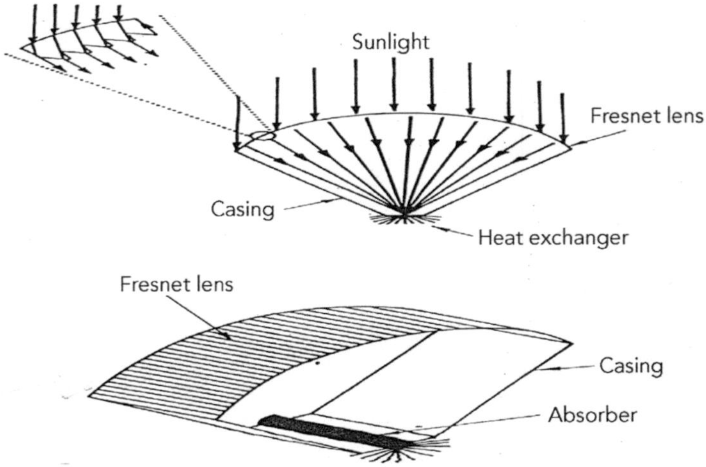

Fig. 9.3 Solar concentrator (Refraction type)

9.3.2 Concentrating Applications

For cost reduction, concentrating module is an important approach to overcome the problem of high cost of solar cell. The low-cost optical concentrator focuses solar radiation onto the cell array of area much smaller than the concentrator aperture (Fig. 9.3). The concentration of radiation also leads to an increase in photon generation current and hence cell efficiencies. The operation of solar cells under high illumination requires the removal of heat from the cells to combat the drop (0.45% of power per degree Kelvin) in efficiencies of cells. The outstanding problems of this technology are the development of

- Efficient low cost concentrator
- Cost-effective tracking systems
- Concentrator cells
- Efficient low cost heat sink

9.4 System Classification

The solar PV power systems have been proposed and tested for generation of electricity in outer space as well as in terrestrial environment. The concept of solar power satellite was proposed in 1964 for the generation of PV electricity in outer space, conversion to microwave of laser beams to be captured and reconverted into ac electricity for

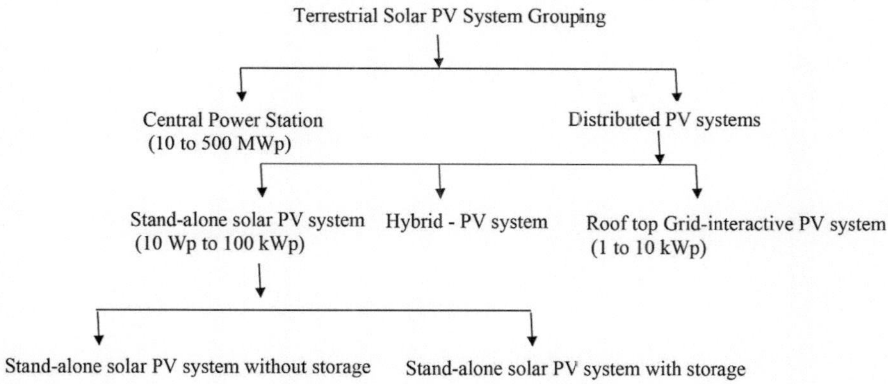

Fig. 9.4 Terrestrial solar PV system grouping with their capacities

distribution to terrestrial load centre. It now appears unattainable even by the end of 2016. However, the use of PV electricity in satellite electronic systems and spacecrafts seems plausible. In terrestrial area the PV electricity has been used in capacity range of 10 Wp–500 Wp. The solar PV system grouping and their capacities range has been shown in Fig. 9.4.

Chapter 10
Solar Photovoltaic System Design

10.1 Solar PV System

The Solar Photovoltaic (SPV) system consists of energy generating subsystem modules, energy storage subsystem battery bank and power conditioning units to supply power, either stand-alone or in conjunction with grid (grid back-up). It is gaining increased importance as a renewable source due to its advantage like absence of fuel cost, little maintenance and no noise and wear due to the absence of the moving parts, etc. In particular, energy conversion from solar cell arrays received considerable attention in the last four decades. Its size is chosen according to planned application. Several demonstration units of large size solar PV power plants have been constructed, operated and monitored around the world. Most of these systems use silicon crystal solar cells. The module may be flat in geometry and use unconcentrated solar irradiance or else it may have concentrating optics (refracting or reflecting) as parts of its structure. Several options are possible in module mounting as follows:

- PV modules may be placed at fixed tilt angle at which the average yearly energy generated is maximized.
- The tilt may be seasonally adjusted; for example at latitude -15, 0 and latitude $+15$ in summer, Equinoxes and winter respectively.
- The modules may be mounted on a tracking unit that follows the sun. Single angle axis trackers follow the sun daily from east to west and two axis trackers further include elevation control to correct for seasonal north-south sun movement.

This is accomplished by a Power Conditioning Unit (PCU), which consists of following sub-systems:

- DC interface
- Charge controller
- Inverter
- Protection circuit
- AC interface

N. D. Kaushika et al., *Solar Photovoltaics*,
https://doi.org/10.1007/978-3-319-72404-1_10

From practical point of view the solar PV system may be divided into following types:

(i) Stand-alone solar PV system without storage
(ii) Stand-alone PV system with storage
(iii) Solar PV-hybrid system
(iv) Grid-Interactive solar PV system

In stand-alone solar PV system without any storage subsystem, PV modules or array supply power to an electrical load (Fig. 10.1). Since the solar PV modules generate variable dc during day time, the load should be chosen in such a way that it should have capability to be operated by variable dc output. The internal resistance of the solar cell depends on incident solar radiation and ambient temperature; the maximum energy will be delivered if internal resistance of the solar cell is equal to the load resistance. Since both the incident solar radiation and ambient temperature vary during the course of a day, it is quite unlikely that internal resistance of solar cell will coincide with the load resistance.

Some power loss occurs due to this mismatch. The power loss due to this mismatch can be overcome by using a maximum power point tracker, which is essentially a dc-to-dc converter (Fig. 10.2a). The converter converts the input dc voltage available from the module to the load voltage without any change in the delivered power, but with a corresponding change in the current. This type of system can only operate as long as enough solar radiation is incident on the module or array. Such a system cannot operate either during night time or whenever there is no solar radiation.

The system can operate an ac load if a dc-to-ac inverter is incorporated between array and the ac load (Fig. 10.2b).

The stand-alone solar PV system with storage battery becomes capable of supplying power during any time of the day and it also provides stable load voltage and power except for water pumping and battery charging applications (Fig. 10.3). A blocking diode is used in order to prevent the reversal of power from battery bank to PV module or array during night time or whenever there is no solar radiation.

In modern sophisticated large capacity stand-alone system, an electronic energy management, control and monitoring subsystems are being used.

A PV-hybrid system is a combination of a stand-alone PV system and either renewable energy sources such as wind, solar thermal or biomass generator or diesel generator (Fig. 10.4). In such a system both the PV and non-PV sources meet the

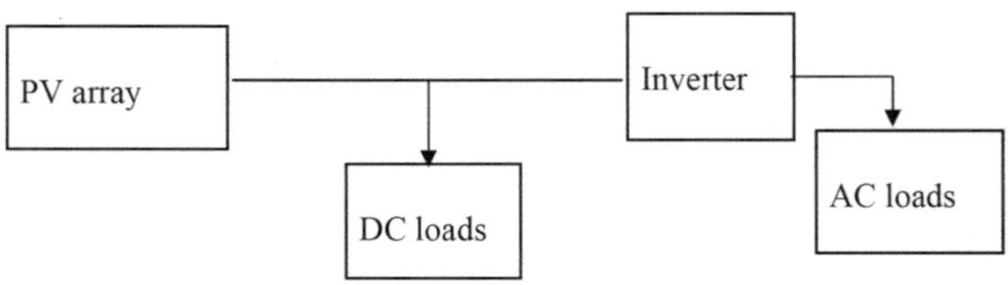

Fig. 10.1 Schematic diagram of solar PV system without battery storage

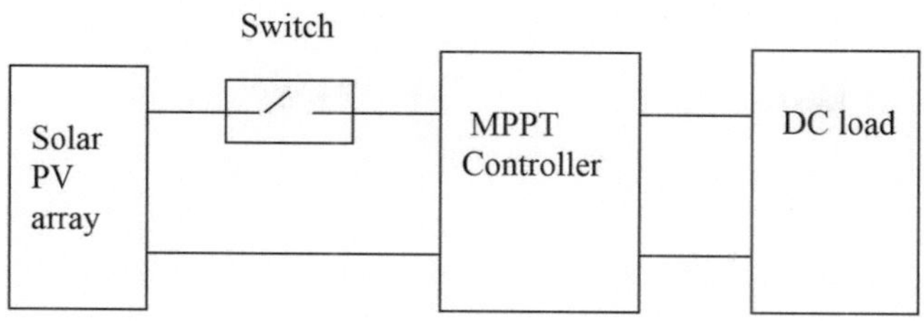

Fig. 10.2a Schematic diagram of solar PV system with MPPT controller for dc load

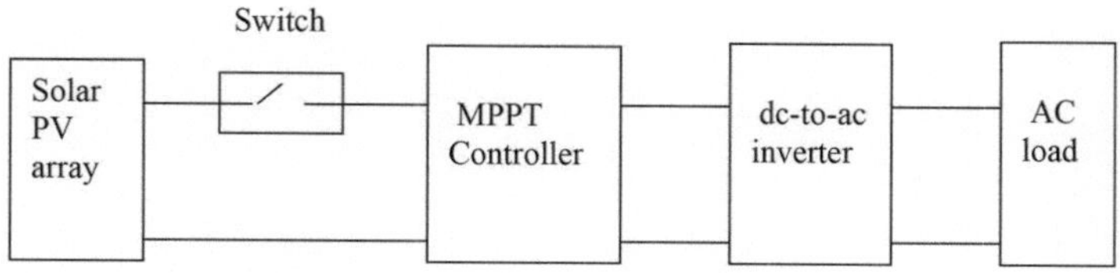

Fig. 10.2b Schematic diagram of solar PV system with MPPT controller for ac load

Fig. 10.3 Schematic diagram of solar PV system with battery storage

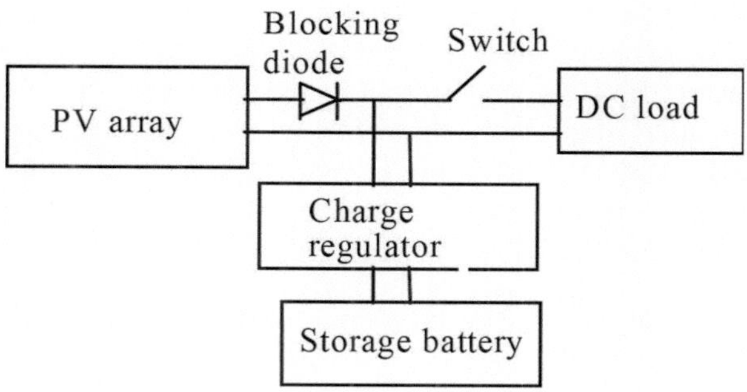

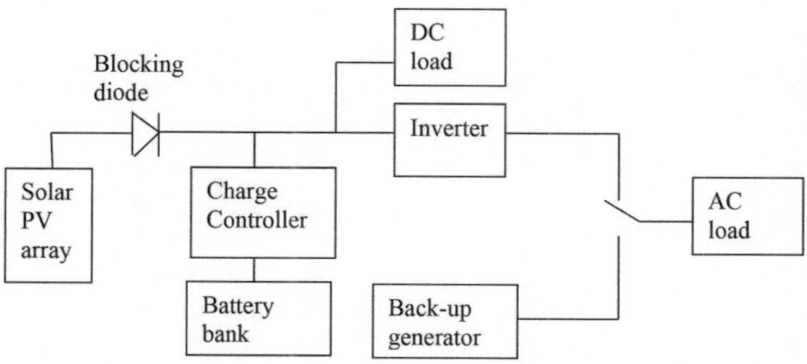

Fig. 10.4 Schematic diagram of solar PV system with battery and back-up generator

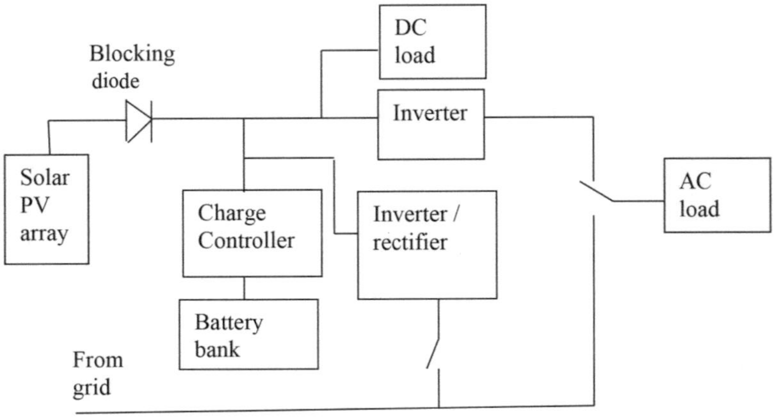

Fig. 10.5 Grid-interactive solar PV system

demand together. In PV-wind hybrid system appear to be a natural mix because both are complementary to each other. The PV energy has been also successfully integrated in hybrid mode with micro-hydel energy sources.

A grid-interactive PV system, solar PV acts only as a supplementary peak-load sharing electrical energy source, whose load becomes the utility grid itself. In grid-interactive PV system an appropriate power condition subsystem, which converts the input dc energy to stable ac energy synchronized to the grid energy, is used for the connection and synchronization of PV array to the utility grid (Fig. 10.5).

The surplus power is fed to the grid through a second inverter, which is usually a line commuted.

10.2 System Design Considerations for Particular Location

Conventional power plants engineering design is based on installed capacity, whereas solar PV system is highly modular in nature; so system capacity is often not a consideration in its design. However, modular capacity addition can be accomplished with due consideration of mismatch losses (Kaushika and Rai 2007). The solar PV system performance depends upon site parameters (i.e. climatological data), system configuration and load parameters. Therefore, it would be desirable to size each PV system individually for each set of site and load. The solar PV systems can be operated in several modes as mentioned in Section 10.1 such as stand-alone with and without battery storage, hybrid and grid connected in accordance with their several applications. The design methods of stand-alone systems only have reached a stage of maturity; therefore, to start, we consider stand-alone solar PV system for a particular location.

10.2.1 Mathematical Formulations

Generally, the PV array is kept inclined at an angle to horizontal, whereas in general the data for incident solar radiation are available corresponding to flat horizontal surface. Following Sukhatme (1984) and Bhattacharya (1998), $G(\beta)$ the factor by which global radiation on a horizontal plane is to be multiplied to obtain that on a surface kept at tilt angle β facing South (or North in Southern hemisphere) is given by:

$$G(\beta) = R(1 - r) + \frac{r}{2}(1 + \cos\beta) + \frac{A}{2}(1 - \cos\beta) \tag{10.1}$$

where

$$R = \left[\frac{\cos(\phi - \beta)\cos\delta\sin\omega_{st} + \omega_{st}\sin(\phi + \omega_{st})\sin\delta}{\cos\phi\cos\delta\sin\omega + \omega\sin\phi\sin\delta}\right]$$

$$\omega_{st} = \cos^{-1}(-\tan(\phi - \beta)\tan\delta) \tag{10.2}$$

$$\omega = \cos^{-1}(-\tan\phi\tan\delta) \tag{10.3}$$

$$\delta = 23.45\sin\left[\frac{360(284 + N)}{365}\right] \tag{10.4}$$

Furthermore, solar array output depends on solar radiation, wind speed and ambient temperature which in turn governs the solar cell temperature. The operating temperature of solar cell can be obtained as follows (Marwali et al. 1997):

$$\begin{aligned} T_c &= T_{amb\cdot} + \left(T_{diff} - \left(T_{gradvw}\right)\right) \\ T_J - T_A &= \text{NOCT} - 20 \\ T_J &= \text{NOCT} + T_A - 20 \end{aligned} \tag{10.5}$$

For Indian climate NOCT is taken as 45 °C (Treble 1991; Kaushika et al. 2005)

$$So \qquad T_J = (T_A + 25) \tag{10.6}$$

The simple and convenient way to express the incident solar radiation on a unit surface area is Equivalent Hours of Full Sunlight (EHFS). It is defined as number of hours of incident sunlight at a place, when the sunlight intensity is constant value of 1000 W/m^2.

To incorporate the effect of cell operating temperature the EHFS is converted into another parameter called UAO, which is expressed in units of Watt hour per peak Watt per day as follows (Bhattcharya 1998)

$$UAO = EHFS \times \left[1 - \frac{\text{Temp.Coefficient in\%}}{100} \times (\text{Monthly max.amb.air temp.})\right] \tag{10.7}$$

PV Array Output Energy = (Peak Power Array capacity at STC) × (UAO) So the energy balance may be expressed as.

PV Array Output Energy = W_L For system without battery storage.

W_L = daily mean load, in W_h (including all losses). For example if W_L' is actual load, then $W_L = W_L'/(f_1 \times f_2 \times f_3)$ where f_1 = field wiring and mismatch loss factor, f_2 = wiring and switch gear loss factor and f_3 = inverter efficiency system with battery storage

$$W_L = \frac{CDB}{n}$$

$$or \quad C = \frac{n\ W_L}{DB} \tag{10.8}$$

where C = capacity in Wh, n = no. of autonomy days, D = maximum depth of discharge and B = battery efficiency.

The first approximation of the array size for a particular location is calculated as follows:

$$\text{Minimum Array Peak Wattage} = \frac{\text{Daily load in Watt hour}}{\text{Daily mean UAO for the year}}$$

Monthly array output is calculated and matched with the load to get the SOC of the battery of the month. If SOC (1-D) is less than the specified value then the array size is increased by 0.1% and the simulation is repeated. Finally the array peak wattage required to maintain the given minimum value of SOC is obtained.

For every location, array is tilted at a fixed angle called optimum array tilt, which is defined as the angle at which the array is tilted from the horizontal plane to receive maximum solar radiation averaged over the year. The optimum array tilt angles will, therefore, be the angle for which the total annual value of Unit Array Output is maximum. Kaushika and Rai (2006) have calculated the optimum tilt angle for 14 stations in Indian region and its variation with latitude is shown in Fig. 10.6 which shows the variation of optimum tilt angle of the array with latitude and fit of second

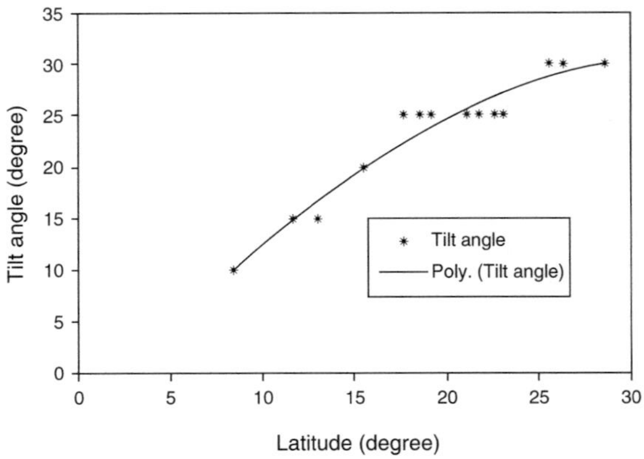

Fig. 10.6 Variation of optimum tilt angle of the array with latitude

degree polynomial therein. The corresponding value of R^2 is also shown; in regression analysis R^2 value is an indicator of the accuracy of best fit. The relationship between optimum tilt angle of the array and latitude of the site under consideration is given by Eq. (10.9).

$$\begin{aligned}\text{Optimum tilt angle} = &-0.0342\,(\text{latitude})^2 \\ &+ 2.265\,(\text{latitude}) - 6.8898 \\ R^2 = &\ 0.95\end{aligned} \tag{10.9}$$

10.3 Solar PV Design Aid Expert System

A solar PV design expert system for the design of solar PV unit for various locations spread over wide region like Indian continent may also be developed. The development of knowledge base is an essential constituent of such an effort.

10.3.1 Development of Knowledge Base for Indian Region

A knowledge base for the design of stand-alone solar PV system corresponding to Indian region for autonomy day of 3 is developed. It considers two modes of system OPERATION: with battery storage and without battery storage. The knowledge base is based on the year round performance of the system corresponding to climatological data (Mani 1980) of 14 stations in Indian region. The knowledge base approach involves combining both site and array characteristics in a single parameter referred to as EUAO and expressing the composite parameter as a function of geographical co-ordinates. A simulation model has been developed for the determination of EUAO for a given load and battery capacity with the main idea that the SOC of the battery does not fall below the specified depth of the discharge during the poor insolation period. The different steps of the mathematical procedure are as follows:

1. Determination of load parameter such as average daily load in Watt-hour (Wh).
2. The battery capacity C is calculated from eq. (10.8).
3. A parameter referred to as EUAO is obtained as follows:

$$\text{EUAO} = \frac{\text{Daily load in Watt hour}}{\text{Required array peak wattage}}$$

Kaushika and Rai (2006) have written a program in C language using the above algorithm. The starting and final results of simulation for New Delhi, India for autonomy days 3 are shown in Tables 10.1a and 10.1b respectively. The variation

Table 10.1a EUAO and array capacity corresponding to climatological data of New Delhi, India for autonomy days 3 and daily mean load 1 kWh in first iteration [Starting: UAO = 5.23 (Wh/Wp/day), Array capacity = 191.05 (Wp)]

At the end of month	Monthly array output (Wh)	Monthly load (Wh)	Surplus/ Deficit (Wh)	Battery SOC (Wh)	Waste energy (Wh)
January	35528.14	31,000	4528.14	4411.76	4528.14
February	33600.61	28,000	5600.61	4411.76	5600.61
March	35981.54	31,000	4981.54	4411.76	4981.54
April	32402.35	30,000	2402.35	4411.76	2402.35
May	31823.15	31,000	823.15	4411.76	823.15
June	25811.09	30,000	−4188.91	222.85	0.00
July	21943.93	31,000	−9056.07	−8833.22	0.00
August	22656.28	31,000	−8343.72	−17176.95	0.00
September	26441.98	30,000	−3558.02	−20734.97	0.00
October	32455.31	31,000	1455.31	−19279.65	0.00
November	32975.26	30,000	2975.26	−16304.40	0.00
December	34024.04	31,000	3024.04	−13280.36	0.00

After first iteration: EUAO = 5.20 (Wh/Wp/day), Array capacity = 192.20 (Wp)

Table 10.1b EUAO and array capacity corresponding to climatological data of New Delhi, India for autonomy days 3 and daily mean load 1 kWh after last iteration

At the end of month	Monthly array output (Wh)	Monthly load (Wh)	Surplus/ Deficit (Wh)	Battery SOC (Wh)	Waste energy (Wh)
January	37627.16	31,000	6627.16	4411.76	6627.16
February	37167.15	28,000	9167.15	4411.76	9167.15
March	42866.29	31,000	11866.29	4411.76	11866.29
April	39734.81	30,000	9734.81	4411.76	9734.81
May	38864.02	31,000	7864.02	4411.76	7864.02
June	33224.18	30,000	3224.18	4411.76	3224.18
July	29211.83	31,000	−1788.17	2623.60	0.00
August	29259.14	31,000	−1740.86	882.74	0.00
September	34772.85	30,000	4772.85	4411.76	1243.82
October	40668.31	31,000	9668.31	4411.76	9668.31
November	39635.66	30,000	9635.66	4411.76	9635.66
December	37861.94	31,000	6861.94	4411.76	6861.94

After last iteration: EUAO = 4.33 (Wh/Wp/day), Array capacity 230.78 (Wp)

of EUAO with latitude and longitude for autonomy days 3 is depicted in Figs. 10.7a and 10.7b respectively. Figure 10.7a shows the variation of EUAO as a function of latitude and fit of polynomial therein. It may be noted that the variation of EUAO with latitude exhibits larger scatter ($R^2 = 0.24$) than the variation with longitude ($R^2 = 0.42$). This

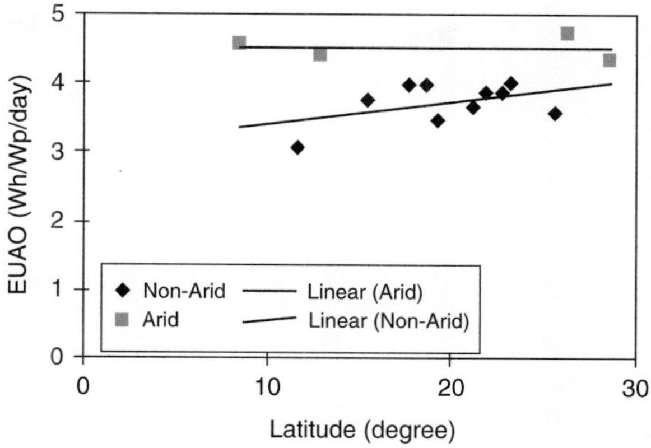

Fig. 10.7a Variation of EUAO with latitude for different longitudes (arid and non-arid zones)

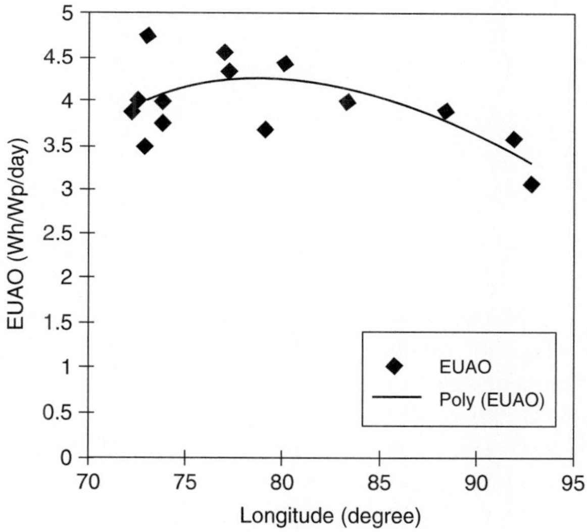

Fig. 10.7b Variation of EUAO with longitude (cumulative station) for autonomy days 3

could be due to specific geometry of Indian Peninsula where the cloudiness index and ambient temperature is controlled by the monsoon. We have, therefore, considered the variation of EUAO with longitude in the composition of our knowledge base. Furthermore, statistically significant polynomial fits are obtained when arid (Latitude 24°–30° and Longitude 68°–75°) and non-arid zones are considered separately (Fig. 10.8).

$$\begin{aligned} \text{EUAO} &= -0.0053(\text{longitude})^2 \\ &\quad +0.8447(\text{longitude}) - 29.344 \\ R^2 &= 0.42 \end{aligned} \tag{10.10}$$

For arid zone

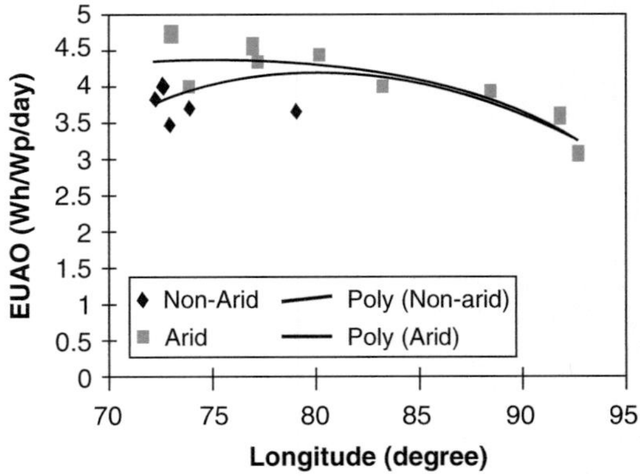

Fig. 10.8 Variation of EUAO with longitude for autonomy days 3 when arid and non-arid zones are considered separately

$$\begin{aligned} \text{EUAO} &= -0.0036(\text{longitude})^2 \\ &\quad +0.5386(\text{longitude}) - 15.783 \\ \text{R}^2 &= 0.79 \end{aligned} \tag{10.11}$$

For non-arid zone

$$\begin{aligned} \text{EUAO} &= -0.0064(\text{longitude})^2 \\ &\quad +1.034\ (\text{longitude}) - 37.413 \\ R^2 &= 0.56 \end{aligned} \tag{10.12}$$

Similarly the variation of EUAO with latitude and longitude for autonomy days 4–7 and the variation of EUAO as a function of latitude and fit of polynomial therein can be obtained.

Kaushika and Rai (2006) developed a software tool referred to as an expert system (Fig. 10.9) for the design of solar PV systems in Indian region. It uses the values of latitude and as well as longitude of location and the number of autonomy days as input and calculates the Effective Unit Array Output (EUAO) from the corresponding polynomial equations. The EUAO is independent of load. For different autonomy EUAO will be different. It then calculates other sizing parameters as follows:

(i) Required array capacity (Wp): It is given by

$$\text{Array capacity} = \frac{\text{Daily} - \text{mean load in kWh}}{\text{EUAO}}$$

(ii) Number of modules in series (N_s): It is given by

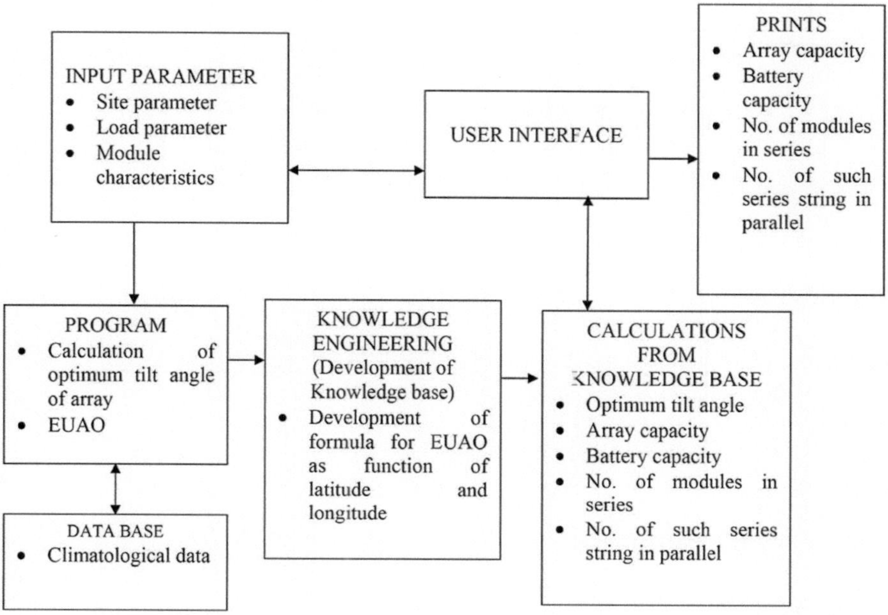

Fig. 10.9 Schematic diagram of an expert system

$$N_s = \frac{\text{Load operating voltage}}{\text{Effective operating voltage of each module}}$$

(iii) Number of series string in parallel (N_p): It is given by

$$N_p = \frac{\text{Required array capacity}}{N_S \times \text{Effective peak power rating of each module}}$$

Where effective peak power rating of each module is effective operating voltage of each module $\times$ I_{max}.

The flow chart of calculations is given in Fig. 10.10.

10.4 Web Access to Solar PV Design Aid Expert System

The website corresponding to the expert system has been prepared and placed on Internet and outline of the uploading procedure is as follows:

(a) We make self-extracting zipped file. The self-extracting zipped file does not require win zip software for extracting the zipped file. The file can be downloaded from anywhere on any machine. The necessity of zip in general

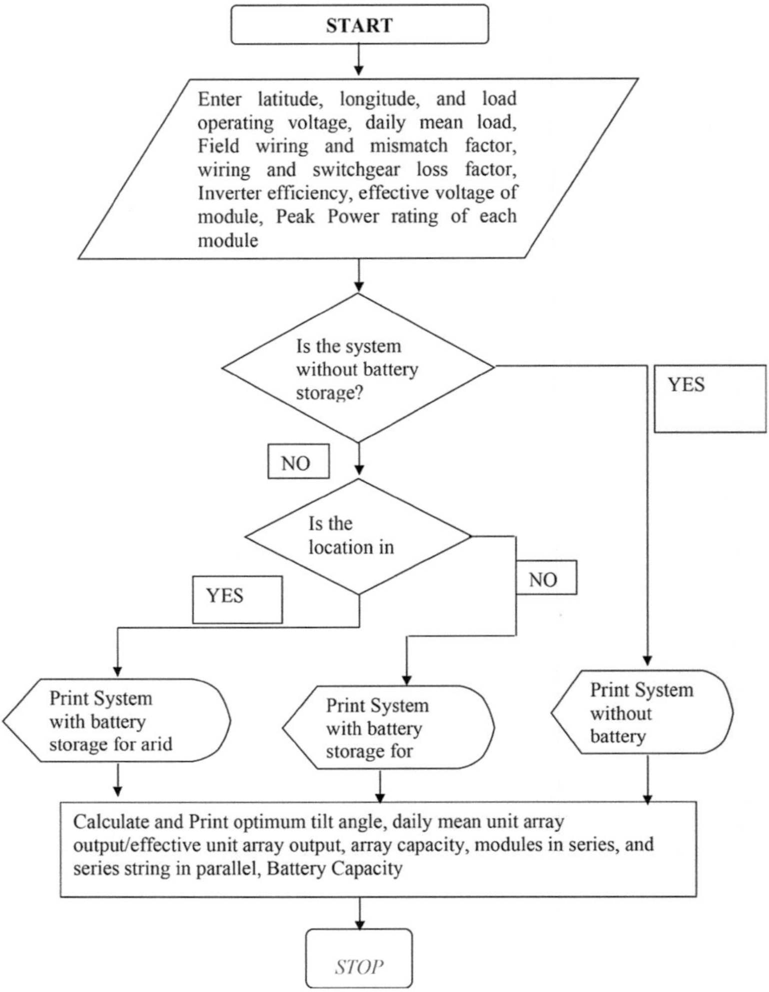

Fig. 10.10 Flow chart for the expert system

arises when there is more than one file accomplishing the required task. In the present case there are more than one file required to run the knowledge-base (e.g. the executive/compiled file of knowledge base program and the executive file of turbo C compiler). The zipping saves space as well as download time.

(b) The executive files of knowledge base and Turbo C are zipped in a single file named setup.exe with the help of WinZip software. Afterwards we make this setup.exe file self-extracting.

(c) Then this setup.exe file is on the web server for downloading.

The address of the websites is http://www.nd-kaushika.tripod.com

Chapter 11
System Reliability Considerations

11.1 The Reliability

The cost and reliability are two outstanding barriers in the promulgation of solar PV technologies on mass scale. In solar PV power system context reliability involves two factors: (i) the solar array need the use of an optimum load to deliver maximum power. This requirement varies with solar irradiance, temperature and battery bank characteristics and often difficult to be realized in practice. (ii) The reliability also means the ability of the system to continue functioning without failure for the period of time intended and under the given operating conditions; for example, during the array operation in field, the partial shadowing, soiling, cracking (provoked by hail impact) or opening of a string may occur due to solder melting or damage to encapsulate and in extreme cases it can lead to system failure. In both cases, the system reliability can only be guaranteed on a probabilistic basis. It essentially involves probability of success/failure, adequate performance, on/off time and operating conditions. Thus the probability provides the numerical input for the assessment of reliability and also the first index of system adequacy. For mission-oriented project this definition of reliability can be regarded as suitable measure. However, it becomes rather unsuitable measure for those continuously operated systems that can accommodate failure. The suitable measure used for these systems is availability. For example, in power system, availability is the percentage of time the system will deliver power to its load (Stember 1981). The required availability is different for different types of systems. Availability may be obtained by dividing "up-time" by "up-time plus down time" or it can be expressed as

$$A = \frac{\text{MTBF}}{\text{MTTR} + \text{MTBF}}$$

where A = availability, MTTR = mean time to repair and MTBF = mean time between failure.

N. D. Kaushika et al., *Solar Photovoltaics*,
https://doi.org/10.1007/978-3-319-72404-1_11

Thus availability can be increased by increasing reliability i.e. making MTBF longer or by decreasing MTTR.

11.2 Site Dependence

The performance of a solar PV system is site dependent since PV array output is a function of cell temperature and solar irradiance which depends on the geographical, climatological and environmental factors like the latitude, global, diffuse and reflected solar radiation, ambient air temperature and wind speed. The array performance also depends on the uniformity and homogeneity of solar irradiance on the PV array surface.

Furthermore at particular location, the angle at which the sunlight is incident on the PV collector surface influences the array output. The performance of the array will, therefore, depend on whether it is made to track the sun during the day, month and year, or it is installed at a fixed tilt angle. A fixed tilt array should face due geographical South in the northern hemisphere and due North in southern hemisphere. The deviation in the azimuth angle from due south (or due north) introduces randomness in system performance. The average daily performance is also dependent on the tilt angle at which the array is installed with respect to the horizontal. The optimum array tilt angle for a fixed-tilt array depends on the local geographical and climatological factors. For example from geographical considerations for maximum yearly energy gain the tilt angle is taken to be $0.9\ L$ where L is latitude angle. For seasonal adjustment the angle to be adjusted $L \pm 15°$. The load as well as system losses do not remain constant over the year. Thus solar PV systems are also highly load-specific.

11.3 System Availability

The system availability means its availability to serve the load; it is an important design parameter for solar PV systems. It is the percentage of time in a whole year, during which the system is able to satisfactorily meet the load. The availability is 100% if the system is able to meet satisfactorily all designated loads during that period. Conventional reliability parameter used in this connection is Loss-of-Load-Probability (LOLP) expressed as the total number of days, during the period of one year, when the system is incapable of meeting the load. The system availability may be expressed in terms of the Loss-of-Load-Probability (LOLP) as follows:

$$\text{System Availability} = \frac{365 - \text{LOLP}}{365} \times 100\%$$

System sizing exercises often attempt to ensure a high value of annual availability or a low value of LOLP. In case of solar PV systems this approach may lead to an expensive system design, because that may demand high system array capacity. On the other hand, a low value of the system availability will normally not be desirable from the viewpoint of the user. Therefore, the system designer must use a pragmatic and judicious approach to ensure reasonably high system availability without making it unduly expensive. An annual availability value between 90% and 99% is generally considered acceptable for most stand-alone systems.

11.4 No Sun Days and Storage Autonomy

The measure of 'sun shine days' has been used as a measure of reliability in solar PV systems. Larger the number of sun shine days higher the reliability of the solar PV system. The number of consecutive off-sun shine days in a month decides the period during which the storage battery is required to meet the load completely. The storage battery provides storage autonomy to counter balance the possibility of no PV generation during off-sun shine (e.g. rainy or cloudy) days. The storage autonomy is referred to as NO-sun days. Larger value of storage autonomy means a larger battery capacity and higher cost.

The capacity C of a storage battery, having storage autonomy of n days, may be expressed as follows:

$$C = \frac{nL}{D \times B}$$

where L is the daily mean value of load, expressed in the same units as the battery capacity C, D is the permitted Maximum Depth of Discharge (MDOD) expressed as fraction and B is the battery efficiency.

11.5 Power Loss and Hot Spot Effects

The undesirable feature of solar PV arrays delivering power lower than the array rating in field conditions could be fundamentally due to mismatch in the output of the individual cells. This power loss may be due to variety of reasons as follows.

11.5.1 Manufacturer's Tolerances in Cell Characteristics

Solar cell fabrication procedure involves several controls despite all automations; the resultant cells possess significant manufacturer's tolerances. Performancewise they exhibit variations in current voltage characteristics. In particular, current (I_{max}) and voltage (V_{max}) of individual cells operating at maximum power (P_{max}) exhibit statistical distribution. So when mass produced cells are connected in series and/or in parallel within a network, the statistical distribution of cell parameter leads to electrical mismatch in the network; the mismatch limits the short circuit current of the combination to a value closer to the lower end of the distribution than to average. Consequently the maximum power output of the network is often less than the sum of the maximum power of individual cell; this is referred to as mismatch loss. It may tend to enhance with time due to degradations resulting from aging. The detailed analysis of power loss due to mismatch in cell characteristics has been carried out in Chapter 7.

Similar results would occur with strings combined in parallel; open circuit voltage distribution will generate inter-string power dissipation.

11.5.2 Environmental Stresses

It includes the partial or full opening of a string due to cell cracking (provoked by hail impact) or otherwise. This not only gives rise to a mismatch loss but also could result in excessive heating of the regions of power dissipation which in turn can lead to solder melting and damage the encapsulant. The regions of excessive heating are referred to as hot spots.

11.5.3 Shadow Problem

In the field solar cells arrays are subject to shadows from both predictable sources as well as from such unpredictable sources as bird droppings or fallen leaves. The percentage power loss is much greater than the percentage of array area which is shadowed; for smaller arrays with few or no parallel connections, one leaf could cause the system output to drop to a fraction of rated power, eventually resulting in system failure. Shadowed cells in series with illuminated cells block the current flow in entire series connection and tend to become reverse biased. Consequently, hot spots are developed in the localized regions of shadowed cell, which can damage the encapsulation or crack the cell causing module failure.

11.5.4 Array Operation at Low Voltage

The operation of the array at low output voltage occurs in some array voltage regulation schemes and is also likely to occur when maintenance is carried out and some sections of the array are sorted out for the safety of the maintenance crew. A low output cell in such cases gets reversed biased and exhibits hot spots. For this reason some manufacturers now specify 'safe operating regions' for their modules.

11.6 System Reliability Improvements

Quality and redundancy are the two main attributes by which the reliability can be affected. Quality concerns not only the physical materials and components used in the system but also the quality of manufacture, testing, calibration, transport and operation. This in turn depends on the quality and experience of personnel involved and the ergonomics and environment of work place. This measure is rather cost-intensive for solar PV system and has been adopted only in such mission-oriented systems as space satellites. The redundancy is based on the basic premise that component will fail from time to time and that there should be sufficient back up in the system so that the function of failed component is absorbed by another. Redundancy is of two types: active and standby. Active redundancy corresponds to the situation that the system components share the normal function amongst themselves; when one or more component fail, the activity is automatically absorbed by the remaining components. This is also referred to as parallel redundancy.

In contradistinction to this is the standby redundancy wherein the redundant component or subsystem remain in a standby state and are switched into the system only when main operating component fails. Reliability improvement requires additional investment and raises a question where or on what the money should be invested in the system to achieve maximum reliability benefit. Reliability and economics, therefore, act as conflicting variables in the decision process. In what follows, therefore, we examine the efficacy of redundancy in terrestrial solar PV system.

The function block diagram of a simple PV system is shown in Fig. 11.1a; it exhibits the subsystem level interconnections. Following Stember (1981) and Billinton and Allan (1992), the above system may be analysed to the subsystem level using two reliability techniques:

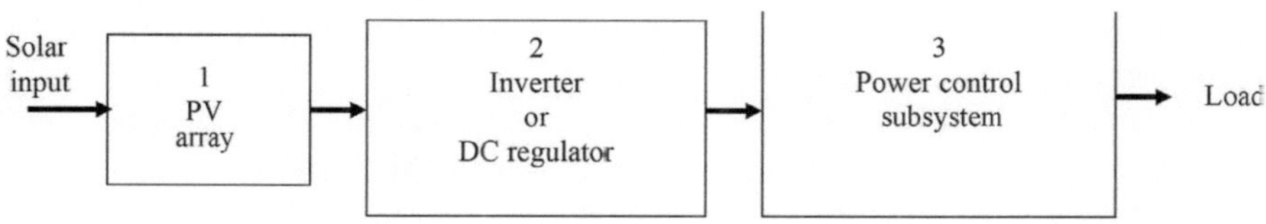

Fig. 11.1a Functional block diagram of a solar PV system.

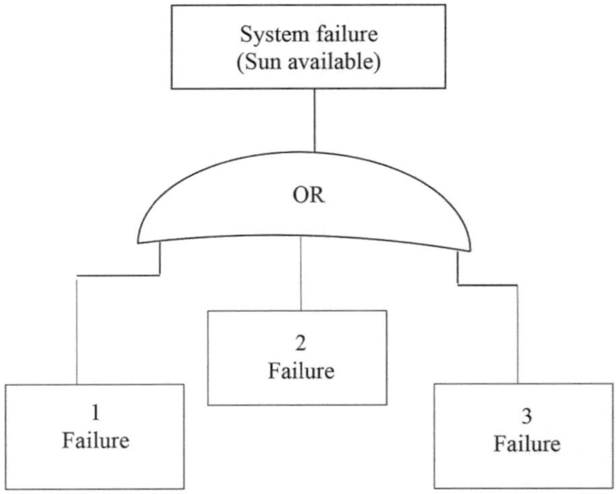

Fig. 11.1b Fault Tree for solar PV system

Table 11.1 Results of fault modes and effects analysis (FMEA) for solar PV system

State of subsystem			
1	2	3	System failure SFNOS
0	0	0	0
0	0	1	1
0	1	0	1
1	0	0	1

0 = Operating, 1 = Failed, SFNOS = System failed no output sun available

- Failure Modes and Effects Analysis (FMEA)
- Fault Tree Model

Fault Tree (Fig. 11.1b) and results of FMEA (Table 11.1) are shown in terms of the consequences of the failure of each subsystem. The system is series connected and an interruption of any of the subsystems will prevent the output being delivered to the load. Consequently, the system is operational only if all the subsystems are operational. Each of the subsystem has a similar effect on the system and its failure causes the failure of the system.

Let us further consider the above system with redundant PV array (Fig. 11.2a). The corresponding FMEA (Table 11.2) and Fault Tree (Fig. 11.2b) show that a failure of either of the PV array does not cause system failure; failure of both only causes the failure. So due to redundant PV array the failure mode is significantly reduced. However, PV array is a costliest component of the system, so its back up redundancy at subsystem level is not an attractive option.

System reliability can also be improved by affecting fault tolerance at component level. In what follows we examine the effectiveness of this approach.

The fault tolerant circuitry can be considered as an arrangement of solar cell interconnections, which is capable of producing acceptable power output even if the

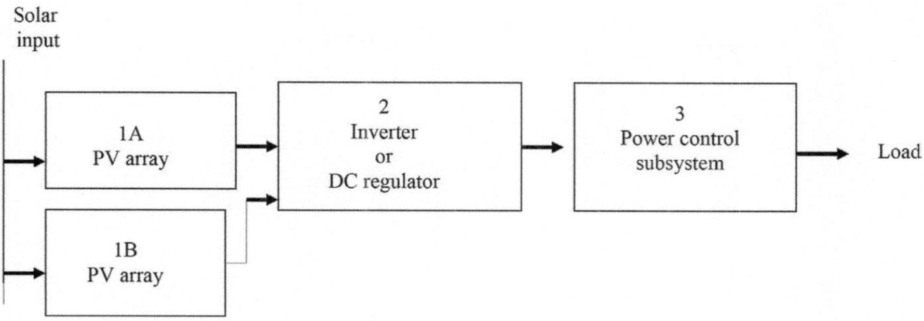

Fig. 11.2a Solar PV system with redundant PV array

Fig. 11.2b Fault tree for solar PV system with redundant PV array

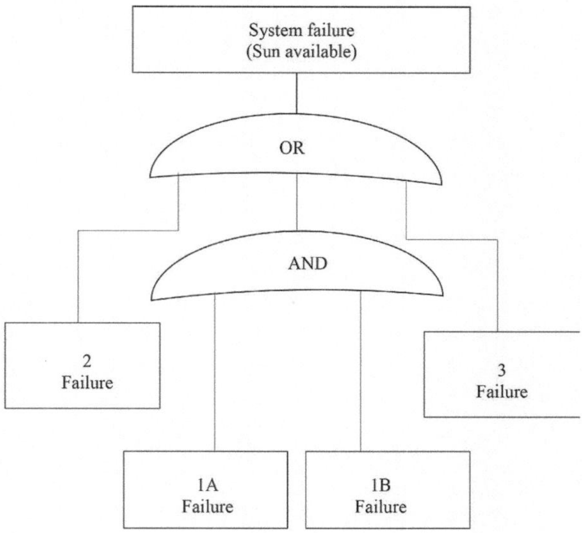

module/array itself suffers from the effects of electrical mismatches. Rauschenbach (1971) proposed and showed that the electrical output of shadowed solar cell arrays could be considerably improved if each series block of parallel string was shunted by a diode.

Subsequently Sayed and Partin (1975) and Swalesh and Green (1982) analysed the effect of shading on terrestrial solar cell arrays and investigated the reduction in power loss and hot spot effects by judicious application of shunt diode. Gonzalez and Weaver (1980) have shown that the efficiency of an array can also be enhanced by redundant circuit design. Green et al. (1981) reported a new technique for integrating the bypass diodes into solar cell structure. Bishop (1988) and Shephard and Sugimura (1984) have studied that the effect of electrical mismatch can be reduced by series-paralleling solar cell interconnection scheme. The effect of electrical mismatches can be substantially reduced and hence the performance of the

Table 11.2 Results of fault modes and effects analysis (FMEA) for solar PV system with redundant PV array

State of subsystem				
1A	1B	2	3	System failure SFNOS
0	0	0	0	0
0	0	0	1	1
0	1	0	0	0
0	0	1	0	1
1	0	0	0	0
0	0	1	1	1
0	1	0	1	1
1	0	0	1	1
0	1	1	0	1
1	0	1	0	1
1	1	0	0	1

0 = Operating, 1 = Failed, SFNOS = System failed no output sun available

array can be improved by a proper arrangement of cross connections in the module. These authors presented experimental results verifying proposed cell processing sequence and system performance of the cells. The concept has the proof but has not yet reached the stage of commercial exploitation. However, most of the present day manufacturers of solar modules use an external bypass diode to take care of partial shadow and opening of a string due to crack in cell. Simultaneously with this practice it has also been realized that one may obtain substantially more output from shadowed array with proper arrangement of cell interconnections in the modules. Several interconnection schemes have been proposed (e.g. Brandstetter 1984; Singer et al. 1984; Gautam and Kaushika 2001; Gautam and Kaushika 2002b). It is of great practical importance to find the configurations less susceptible to hot spots and excessive power loss. In what follows we investigate four configurations of array:

- Series-paralleling (SP) configuration
- Series parallel with bypass diode
- Bridge-linked (BL) configuration
- Partially cross-ties configuration

The schematic of series-paralleling concept is illustrated in Fig. 11.3a. To ascertain the effectiveness of various degree of series paralleling, the power output of an array under specific failure condition must be determined for variety of series-parallel configuration. The power loss varies inversely with degree of series paralleling, but limit imposed by physical constraints, as well as cost and thermal problem, make it impossible to achieve negligible power loss by means of unlimited series paralleling.

Figure 11.3b exhibits the use of bypass diode. A bypass diode is connected in parallel with a given number of series blocks. The bypass diode does not conduct unless the series block is reversed from normal polarity by a small amount, usually say 0.5 V. Then the diode passes unlimited current at its operating voltage. The

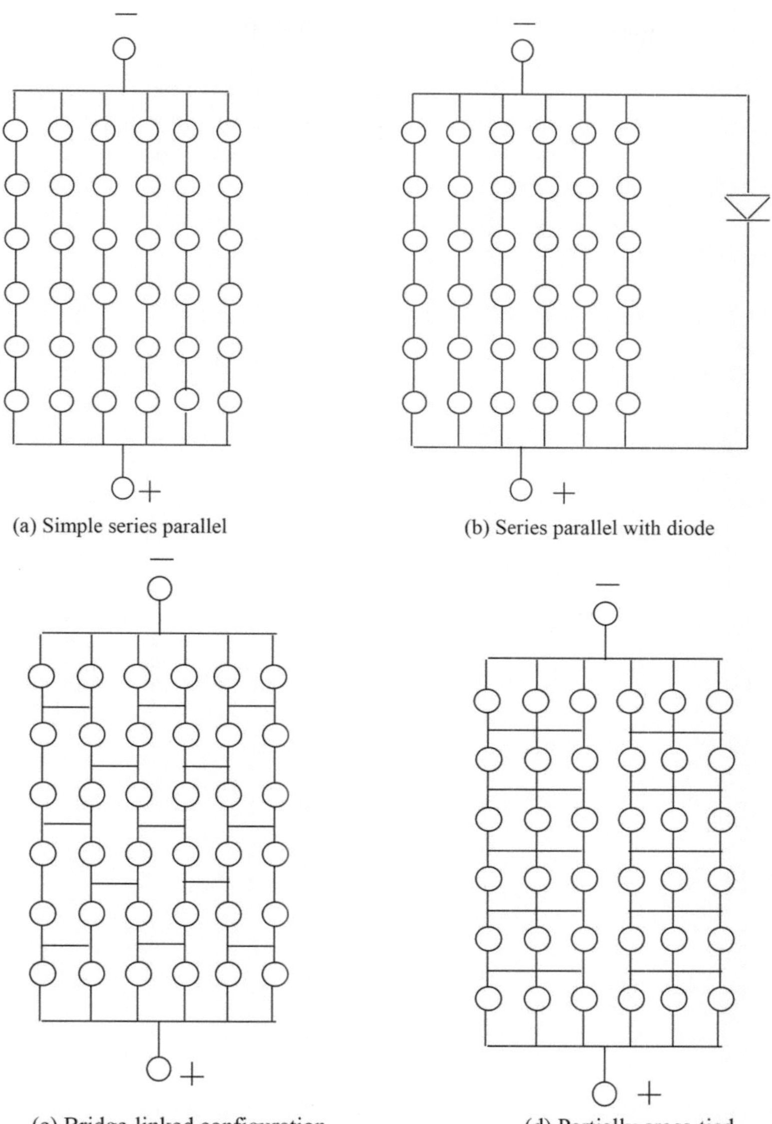

(a) Simple series parallel

(b) Series parallel with diode

(c) Bridge-linked configuration

(d) Partially cross-tied

Fig. 11.3 Different interconnection schemes for solar PV array

purpose of using a diode is to prevent the series block from operating at high reverse voltage. In this state it absorbs power from the system and cells operating with reversed polarity becomes hot, leading to physical degradation of cell and module. This condition is known as back biasing which results in hot spot problem.

Figure 11.3c illustrates bridge-linked configuration, which is accomplished by making cross connections between parallel strings of cells of series-parallel configuration. The cross connections are in bridge rectifier fashion so called as bridge-linked configuration. The power loss varies inversely proportional to square root of M (parallel string) $\times N$ (series block) so the relative advantage of the BL array over the SP array increases with increasing the array size. In the BL array the surrounding cells share the effect of reverse bias on the shadowed cell.

Figure 11.3d shows partially cross ties, which is obtained from the simple series-parallel array by connecting ties across row of junctions; it may be characterized as the scheme with higher cell interconnection redundancy. The experimental investigations of electrical mismatches due to shadow effect have been carried out to find an array configuration, which is comparatively more fault tolerant. Module of size 6 $\times$ 6 (Fig. 11.3) have been considered to carry out the investigations. Two cells of a string were shadowed. Figures 11.4, 11.5 and 11.6 portray corresponding I-V curves for the

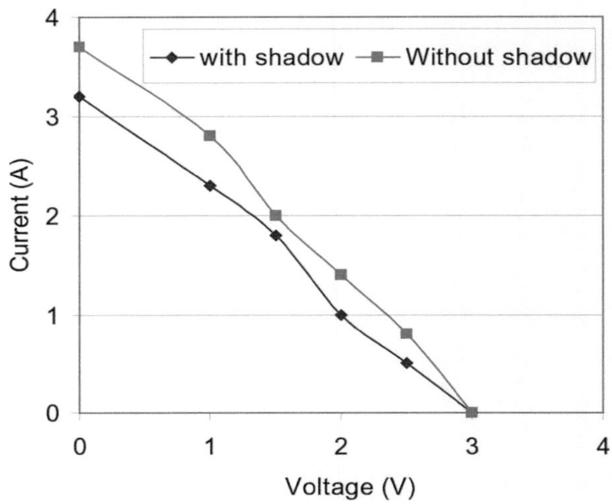

Fig. 11.4 I-V characteristics when solar cells are connected in series-parallel with diode

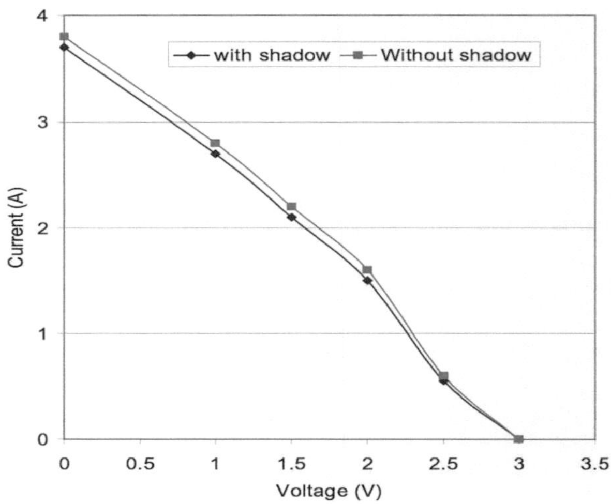

Fig. 11.5 I-V characteristics when solar cells are connected in bridge-linked configuration

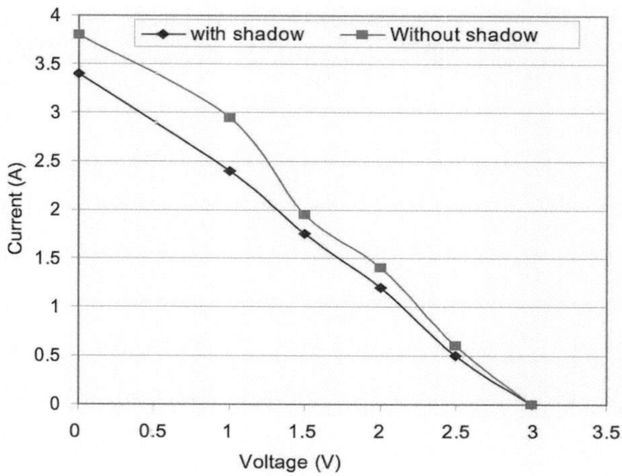

Fig. 11.6 I-V characteristics when solar cells are connected in partially cross-tied configuration

interconnection schemes of Fig. 11.3. These are drawn under conditions of known temperature and stable solar irradiance. The results correspond to 6 × 6 modules wherein two cells of a string have been shadowed. It is obvious that the bridge linked solar cell interconnection configuration turns out to be superior to both series parallel and partial cross ties configurations in fault tolerance due to shadow effect.

Chapter 12
High Performance Solar Cells

12.1 Solar Cell Developments

In recent years research in solar cell physics and reliability aspects of solar photovoltaic systems has exhibited potential for significant cost reduction as well as functionality amelioration. In this chapter high efficiency and cost-effective approaches in solar energy conversion to make solar photovoltaics a commercial perspective has been discussed.

Many solar cell technologies using different types of materials are available today, and an even larger number is expected to be available in the near future. These may be classified into three generations, depending on the basic material used and the level of commercial maturity. The basis of classification is as follows:

First-generation PV systems (fully commercial) use the wafer-based crystalline silicon technology, either single crystalline or multi-crystalline.

Second-generation PV systems are based on thin-film PV technologies and generally include three main families: (1) amorphous (a-Si) and micromorph silicon (a-Si/μc-Si); (2) Cadmium-Telluride (CdTe); and (3) Copper Indium-Selenide (CIS) and Copper-IndiumGallium-Diselenide (CIGS).

Third-generation PV systems include technologies, such as advanced thin film technologies (multi junction technologies), concentrating PV (CPV) technologies and organic PV cells technologies that are still under demonstration. It also includes novel concepts under development.

12.2 First Generation Solar Cells

It corresponds to fully commercial solar cells which include cells made from pure mono-crystalline silicon with almost no defects or impurities. These cells in general have a solar conversion efficiency of about 15–17%. The most efficient and

N. D. Kaushika et al., *Solar Photovoltaics*,
https://doi.org/10.1007/978-3-319-72404-1_12

expensive single-junction silicon cells have been reported to have about 22% efficiency. The fabrication of these devices is energy, material and time intensive; they are expensive to produce. As a follow up to these cells, the polycrystalline solar cells were produced from slightly poorer grades of mono-crystalline silicon or semiconductor grade silicon. The cells have white speckles on the surface due to impurities. They are comparatively less expensive since simpler processes are involved in their production. They have a solar conversion efficiency of about 10–12%.

12.3 Second Generation Solar Cells

Cost of electricity generation from solar cells depends on many factors but most important are the cost of material used in the fabrication of solar cells and the cost of manufacturing processes. The reduction of these costs is achieved in solar cell made from thin film of material rather than from usual slice of crystalline silicon. The thickness of the thin film can range between 10–100 micron (10^{-6} m); there are wide variety of techniques which can be used to deposit thin films of amorphous silicon as well as of several polycrystalline materials (the materials are intermediate between amorphous and crystalline). They absorb sunlight more effectively than crystalline cells and can therefore be thinner. Amorphous silicon solar cells have lower manufacturing cost than crystalline cell but have a low conversion efficiency of 5–7%. However the reduced cost often overweighs the reduced efficiency leading to the increase in ratio of performance to cost. Almost all polycrystalline film solar cells are heterogeneous consisting of an *n*-type material and a different *p*-type material. Two types of polycrystalline thin film cells have been shown to have characteristics which make them attractive for commercial production. These are: the copper indium diselenide/cadmium sulphide cell and cadmium telluride/cadmium sulphide cell. Following three primary types of thin film solar cells have been commercially developed:

1. Amorphous silicon (a-Si) and microcrystalline silicon (micro-Si);
2. Cadmium Telluride (Cd-Te); and
3. Copper-Indium-Selenide (CIS) and Copper Indium-Gallium-Diselenide (CIGS).

Thin film technology has been successfully used on rigid, flexible, curved and foldable substrates. Amorphous silicon efficiencies are rather low 4–8% and amorphous silicon with layer of micro crystalline silicon may increase the efficiency up to 10%. The Cadmium-Telluride thin film cells have lower production cost and higher efficiencies up to 16.7% (Green et al. 2011). However, the main disadvantages are that cadmium is byproduct of zinc mining; tellurium is byproduct of copper processing and cadmium has issues related to its toxicity. CIS and CIGS have the highest efficiencies of all thin film cells and efficiency of 23.3% has been achieved in the laboratory.

Multiple junction approach has also been proposed (Jackson 1955) for the fabrication of high performance solar cells. The approach is depicted in Chambouleyron (1991). The multiple gap devices consists of a series of single gap cells fabricated one over the other; each one uses a different region of solar spectrum. In principle, multijunction devices with any of single crystal, poly crystalline or amorphous structure can be built. However, the single crystal structures involve the problem of lattice mismatch between different materials. Current research addresses such problems.

12.4 Third Generation Solar Cells

Over the years the cost of solar photovoltaic electricity has gradually decreased but it is still far from proving to be affordable by masses. Third generation solar PV systems are characterized mainly by the fact that they offer substantial improvement in the efficiency of solar conversion or a large reduction in their cost of production as compared to previous generation of system.

For example, in the last few years there has been an interest in thin film nano crystalline silicon solar cells. Low temperature plasma enhanced chemical vapour deposition (PECVD) has been considered as a suitable deposition method (Ray 2009). In university of Neuchatel Switzerland, Shah et al. (2003) have fabricated nanocrytalline silicon solar cell using very high frequency glow discharge method and attained 7% efficiency. Yet another class of thin film solar cells includes organic solar cells (Pulfrey 1978; Hall et al. 2001) and dye-sensitized solar cell (Brian O'Regan and Gratzel 1991). The potential of organic cells to be a viable candidate for solar electricity generation hinges on various factors such as materials, production techniques, working principles and the stability of solar cell device. Gunes et al. (2007) have reviewed these aspects of organic solar cells. Materials having delocalized π electron system can absorb sunlight, create excitons and transport these carriers. The organic solar cells use conductive organic polymers or small organic molecules materials for light aborption and production of excitons (Suresh et al. 2009; Sharma 2009; Lokhande and Dhawla 2009).

The effective field for breaking up of excitons into free electron-hole pair is set up by the heterojunction resulting from donor (D) and acceptor (A) material. Two types of architectures namely, bilayer heterojunction (Hoppe and Sariciftci 2004) and bulk heterojunction (BHJ) have been proposed and tested (Sharma 2009). BHJ solar cells have essentially used soluble conjugate polymers and small molecules as donor materials. The solar conversion efficiency up to 7.4% has been reported for these cells (Mikroyannidis et al. 2009; Sharma et al. 2010). Dye-sensitized solar cell (DSSC) is based on the basic premise that organic dyes can generate electricity at oxide electrodes and it is basically a photoelectron chemical system (Brian O'Regan and Gratzel 1991). It is often referred to as Gratzel cells. In dye-sensitized solar cell the photoelectrons are provided from the photosensitive dye and the charge separation occurs at surfaces between the dye semiconductor and electrolyte (Sivakumar and Anandau 2009). Its function is based on photo induced charge separation by dye-sensitized interface between a nano crystalline, mesoporous metal oxide electrode and redo electrolyte. Overall peak power conversion efficiency for the dye-sensitized solar cell has been reported 11% (Gao et al. 2008) and record efficiency for prototype lies at 12.3% (Yella et al. 2011).

Furthermore, the major barrier is the low conversion efficiency of first and second generation solar cells. The most efficient and expensive single-junction silicon cells have been reported to have about 22% efficiency and the upper obtainable efficiency limit (Shockley-Queasier) for a perfect silicon homojunction solar cell is 31%. It is due to inefficient utilization of solar radiation spectrum and single electron band gap of solar cell material. The definition of third generation solar cells has, therefore, been set as the solar conversion efficiency greater than theoretical efficiency (Shockley-Queasier limit) and low cost. The main approaches include multi junction cells, quantum dot technologies and modification of the photonic energy distribution prior to absorption in the solar cell.

In this category of high efficiency solar cells nanostructure solar cells have been reported which include several advanced concepts of quantum dot solar cells (e.g. Nozik 2001). Quantum dot is a special type of nanocrystal. It enables several advanced concepts of solar cell design to be implemented which includes intermediate band and multiple exciton devices, hot carrier and up/down conversion devices.

12.5 Limited Demonstration of Concentrating Solar Photovoltaic

We have come across limited demonstration of concentrating solar photovoltaic system. This includes reflecting concentrating systems by Tata BP Solar India and concentrating solar photovoltaic refracting type 300 kW and 20 kW Entech Solar PV Systems, USA (Figs 12.1a, b).

(a) (b)

Fig. 12.1 Concentrating solar PV system: reflecting type. (**a**) 300 kW solar PV system. (**b**) 20 kW solar PV system

Chapter 13
Solar PV System Economics

13.1 General

In the present day world we come across various energy supplies or projects which are based on many energy sources (conventional and new as well as renewable). Consequently, there are various forms of utilizable energy, various utilization technologies, and various applications at various locations. Presently solar photovoltaics is being recognized as a principal emerging technology growing well over 30% every year. Mass scale adoption of solar PV technology as an energy resource will depend on how this is worth economically. Operationally the task of evaluation consists of identifying the potential advantages and disadvantages of this option. This constitutes the problem of system evaluation.

13.2 System Evaluation

The methodologies of system evaluation have been developed in three disciplines as follows:

1. Engineering: Benefit-cost or cost-benefit analysis
2. Economics: Social and welfare economics or social cost-benefit analysis
3. Operations research: Decision analysis

Each of these disciplines focuses on a separate set of issues. Each is, therefore, appropriate for a different kind of problem. So the first task of an analyst is to choose the best method for a specific problem. The second task involves 'precision' in decision making. It depends on the assumptions one makes about the situation. As a rule precision is decreased by making more simplifying assumptions; conversely, more simplifying assumptions make the evaluation less acceptable.

N. D. Kaushika et al., *Solar Photovoltaics*,
https://doi.org/10.1007/978-3-319-72404-1_13

To illustrate the relationship between simplification, precision and reality let us consider an example for evaluation. Suppose we have savings and we want to invest, following choices are available for different interest rates as follows:

Choices	Different monetary return (in terms of interests)
Saving bank account	7%
Fixed deposits receipt (with bank)	10%
Fixed deposit receipt (Public sector)	14%
Fixed deposit receipt (Private sector)	18%
Investment in business	30%

In the above situation the decision making involves the following complicating factors:

1. Risk
2. Our objective
   (a) Make profit only
   (b) Protects savings
   (c) Flexibility in use

If we consider only the monetary gain the decision becomes pin pointed but it involves elimination of risk and flexibility to use the capital. If we further eliminate the consideration of risk the decision becomes more unrealistic for the results to be acceptable. Further complicating factor is flexibility in use.

13.3 Time Value of the Money

The most basic approach to evaluation in engineering economics is to encompass the situation wherein one considers comparing money over large span of time. For such situations in which various projects have predominantly financial effects and investment at different times, it is, therefore, suitable either to reduce cost or produce marketable goods at same time.

Furthermore, the engineering projects evolve over a long time. Costs incurred in one period generate benefit for many years. The essential problem in the evaluation of such projects comes from the fact that money has a time value. A dollar now and a dollar later are not the same. We cannot estimate total benefit (costs) simply by adding up the benefits (or costs) that occur at different times. This would be like adding apples and oranges. We, therefore, need to translate all costs and benefit with comparable amounts. This is by the use of parameter known as discount rate which may be defined as follows:

- A dollar now is worth more than a dollar in future because it can be used productively between now and then. For example, one can put the money in saving bank account and get a large amount back.

- Furthermore, any given amount of money now is typically worth more than the same amount in future because of inflation. As prices go up due to inflation, the money you have now will buy less and less.

The discount rate represents the way money now is worth more than money later. It determines by how much any future amount is discounted or reduced to make it correspond to an equal amount today. Discount rate is similar to interest rate but with a difference that discount rate represents real change in value to a person or group as determined by their possibilities for production use of money and the effects of inflation. As a general rule, discount rate is different than interest rate. For the time being, it is sufficient to recognize that the discount rate is mathematically similar to an interest rate and is related to yet another term escalation ratio, d through the escalation rate i.

The discount rate appears in evaluation calculations as the discount multiplier $(1 + r)$ where r is a fraction. The influence of currency inflation and other commodities market forces is generally exposed by a fraction referred to as escalation rate. The escalation rate appears in evaluation calculation as the escalation multiplier $(1 + e)$ where e is a fraction.

The rate of escalation multiplier $(1 + e)$ to discount multiplier $(1 + r)$ is referred to as escalation ratio, d and given by

$$d = \frac{1 + e}{1 + r}$$

13.4 Formulations for Evaluation of Money Value Over Time Span

13.4.1 Notations

Present amount, P: It is either money spent or received now, or the value now of future sums.

Future amount, F: It is some amount in future, N, periods from now.

The series of equal amount, R: It is a constant stream of equal amounts received at the end of each of N periods.

13.4.2 Period

It is the fixed interval of time for which the discount rate is defined. It is typically but not necessarily a year.

1. *The Compound Amount Formula*
   Future value F a given amount P at present after N periods is given by

$$F = P(1 + r)^N \quad (r = 0.1 \text{ for } 10\%)$$

Compound amount factor = (1 + r)N

2. *The Present Value Formula:*

It is present value P of a future sum at N years. It is inverse of F.

$$P = F(1 + r)^N$$
$$\text{Present value factor} = (1 + r)^N$$

3. *Capital Recovery Formula*
   It defines the stream of N constant payments that are equivalent to a present sum, P.
   The value of future sum that is equal to the series is given by

$$F = \sum_i^N R(1 + r)^i = \frac{R\left[(1 + r)^N - 1\right]}{(1 + r) - 1} \quad \underset{1}{\uparrow} \quad \frac{\uparrow \text{---------} \uparrow \text{--------} \uparrow}{\quad\quad\quad\quad\; \text{i} \quad\quad \text{N}}$$

The equivalent present value is obtained by dividing by (1 + r)N

As $$P = \frac{F}{(1 + r)^N}$$

So after rearranging we have

$$R = \frac{Pr(1 + r)^N}{\left[(1 + r)^N - 1\right]}$$

Capital recovery factor, c r f, is given by

$$R = P(\text{c r f}) \text{ and c r f} = \frac{r(1 + r)^N}{\left[(1 + r)^N - 1\right]}$$

If (1 + r)N > > 1 as N is large c r f → r

4. *Series Present Value Formula*

$$P = R(\mathrm{c\,r\,f})^{-1} \text{ Series present value factor} = \frac{1}{\mathrm{c\,r\,f}}$$

13.5 Cost-Benefit Analysis in Engineering Systems/Projects

The cost-benefit analysis in engineering systems is used for:

1. Identification of worthwhile projects
2. Ranking of projects.

The methodology involves the use of an index of merit based on accumulated costs and benefits. In practice there are many indices and each one is used in accordance with the nature of problem. These indices may be classified into two categories.

13.5.1 Category I

It is based on the variety of ways of using discounted cash flow of benefits and costs associated with any project. These are the net present value, the benefit-cost ratio and internal rate of return.

13.5.1.1 Net Present Value (NPV)

The NPV of a project is simply the difference between the discounted benefits B and cost C associated with the project:

$$\mathrm{NPV} = B - C$$

Advantages

The most important advantage of NPV criteria is that it focuses attention on quantity of money. Such an advantage is not available with methods which rank projects by ratios.

Disadvantages

1. At the first glance NPV suggests profit. Let us see whether it really implies profit. In ordinary sense profit is difference between what we receive and what we pay out. Consider an example: an investment now for a lump sum of revenue later. So

   Profit = Money received − Money invested.
   Profit = Money received − Money invested $(1 + i)^N$

where i is interest rate.

Present value of profit = (Money received) $(1 + r)^{-N}$ – (Money invested) $(1 + r)^{N}$ $(1 + r)^{-N}$

So Net Present Value is not equal to present value of profit, because discount

2. Secondly, it gives no indication of the scale of effort required to achieve the result. For example see the table below where NPV criterion marks the scale of effort required:

Project	Benefits (Rs.)	Cost (Rs.)	Net value (Rs.)	NPV % of cost
S	2002,000	2000,000	2000	0.1
T	2000	1000	1000	100

So, net present value method by itself is not a good criterion for ranking projects.

The essential conditions for NPV methods are:

We have a fixed budget to invest.

Projects require the same investment.

13.5.1.2 Benefit-Cost (Cost-Benefit Ratio)

Benefit–cost or cost-benefit ratio is the quotient of monetary value of benefits of a project divided by its costs.

$$\frac{Benefit}{Cost} = \frac{B}{C}$$

Advantages

- It compares projects on a common scale.
- It directly provides an indication of whether a project is worthwhile.
- It provides an easy means to rank projects in order of relative merit.

The major analytic weakness of the benefit-cost ratio lies with the ambiguity of the treatment of recurring costs and consequent bias in favour of capital intensive projects. This is illustrated in Table 13.1.

Benefit/Cost ratio:	1.34 for K-project is better than 1.17 for R-project
Annual return:	15% for K-project is worse than 20% for R-project
NPV:	Rs. 500,000 for K-project is worse than Rs. 1000,000 for R-project

Table 13.1 A comparison of capital intensive and operations projects (All costs and benefits at present value).

Project	K-project	R-project
Investment (C_k)	Rs. 1000,000	Rs. 1000,000
Annual cost (C_r)	Rs. 50,000	Rs. 500,000
Annual benefits	Rs. 200,000	Rs. 700,000
Annual net return	Rs. 150,000	Rs. 200,000
Usual life	10 years	10 years
Total benefits	Rs. 2000,000	Rs. 7000,000
Total cost ($C_k + C_r$)	Rs. 1500,000	Rs. 6000,000

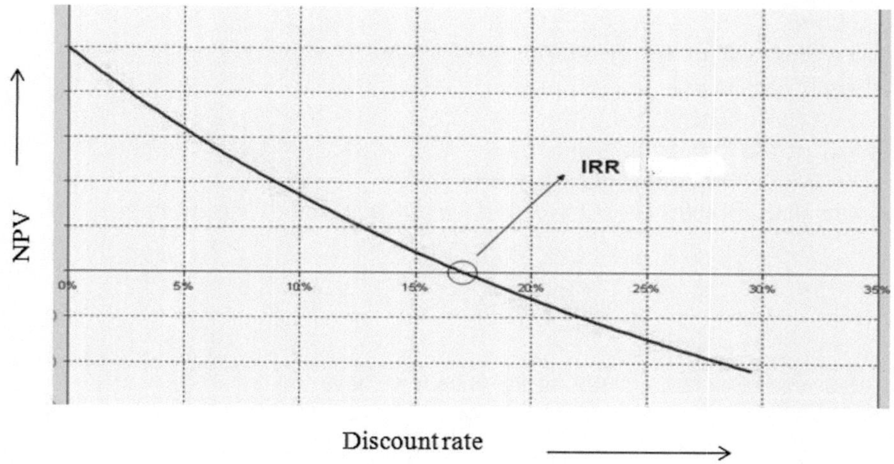

Fig. 13.1 NPV as a function of discount rate.

13.5.1.3 Internal Rate of Return (IRR)

It is the discount rate for which the NPV is equal to zero and the project should be thus ranked from the highest to lowest internal rate of return.

Advantage: No necessity to find the discount rate. Instead one can plot NPV as a function of discount rate (Fig. 13.1) and determine IRR.

13.5.2 Category II

It has been devised to circumvent the problems in category I. It includes cost effectiveness ratio payback period.

Payback period may be defined as the number of years required for the capital to be recovered from the net cash flow of benefits. For example, in solar PV systems saving of conventional energy accrue from the use of solar energy in solar PV system.

13.5.2.1 Mathematical Formulation of Cost-Benefit Analysis

Let us consider a system involving an initial capital *I* which saves energy *S*. The present cost of this energy be C_0 per unit of time and let this cost of energy escalate *e* per unit of time; then let the initial capital investment be *I*.Then at the end of *N* units of time

The future value of the investment = *I* (1 + *r*)*N*

As energy saved = *S*

Present cost of energy = C_0

At any period $C_i = C_0\ (1 + e)i$

where *e* is escalation rate of energy cost.

Accumulated benefits (*B*) at the end of the payback period, *N* is given by

$$B = \sum_{i=1}^{N} S C_i\ (1+r)^{N-i}$$

For *N* to be payback period of the system we have

$$I(1+r)^N = C_0 S (1+r)^N \sum_{i=1}^{N} \left(\frac{1+e}{1+r}\right)^i$$

So,

$$\frac{1}{C_0 S} = \sum_{i=1}^{N} \left(\frac{1+e}{1+r}\right)^i$$

or,

$$\frac{I}{C_0 S} = \frac{\left(\frac{1+e}{1+r}\right)\left[\left(\frac{1+e}{1+r}\right)^N - 1\right]}{\frac{1+e}{1+r} - 1} \tag{13.1}$$

If we put

$$\frac{r-e}{1+e} = W$$

or

$$1 + W = \frac{r + 1}{1 + e}$$

$\therefore$

$$\frac{1}{C_0 S} = \frac{\frac{1}{1+W}\left[\left(\frac{1}{1+W}\right)^N - 1\right]}{\frac{1}{1+W} - 1}$$

$\therefore$

$$\frac{I}{C_0 S} = \frac{1}{W}\left[1 - \left(\frac{1}{1 + W}\right)^N\right]$$

$$\left(1 - \frac{WI}{C_0\, S}\right) = \left(\frac{1}{1+W}\right)^N$$

or,

$$-N\log(1 + W) = \log\left(1 - \frac{WI}{C_0\, S}\right)$$

or,

$$N = \frac{-\log\left(1 - \frac{W}{ROI}\right)}{\log(1 + W)}$$

where

$$ROI = \frac{C_0\, S}{I}$$

As $C_0 S$ is return on investment in first unit of time and I is initial investment. Alternatively, the Eq. (13.1) can be expressed as

$$\frac{1}{C_0 S} = \frac{\left(\frac{1+e}{1+r}\right)\left[\left(\frac{1+e}{1+r}\right)^N - 1\right]}{\frac{1+e}{1+r} - 1}$$

Let

$$d = \frac{1 + e}{1 + r}$$

$$\frac{I}{C_0 S} = \frac{d\left[(d)^N - 1\right]}{d - 1}$$

or,

$$\frac{I}{C_0 S}\frac{d-1}{d} = (d)^N - 1$$

or,

$$\left\{\frac{I}{C_0 S}\frac{d-1}{d}\right\} + 1 = (d)^N$$

or,

$$N \log d = \log\left[\left\{\left(\frac{I}{C_0 S}\right)\left(\frac{d-1}{d}\right)\right\} + 1\right]$$

or,

$$N = \frac{\log\left[\left\{\left(\frac{I}{C_0 S}\right)\left(\frac{d-1}{d}\right)\right\} + 1\right]}{\log d}$$

In practice the life cycle costing is often used for determining the most cost-effective energy solution. In the life cycle cost, there are various levels of sophistication and accounting ranging from payback period to net present cost for the project. There are several assumptions such as future values of discount rate and the energy price escalation as well as several formats for expressing the figure of merit for decision making. In what follows we present some illustrative numerical examples.

13.5.2.2 Problem

A roof top integrated solar PV system 100 kWp capacity is installed at Delhi characterized by solar insolation of 4.2 kWh per day. The power generated is injected into Low Tension (LT) electricity network. The power cost obtainable from the grid owner is Rs. 8.00 with 10% escalation every year. The initial cost is Rs. 70 lacs. Calculate the payback period of the system.

Using the expression for payback period, N

$$N = \frac{-\log\left(1 - \frac{W}{ROI}\right)}{\log(1 + W)}$$

Here,

$$ROI = \frac{C_0 S}{I}$$

From the given problem

$$I = 70\,\text{lacs}$$
$$C_0 S = 4.2 \times 100 \times 360 \times 8 = 1209,600$$
$$ROI = \frac{1209,600}{7000,000} = 0.1728$$
$$W = 0.1$$
$$1 - \frac{W}{ROI} = 1 - \frac{0.1}{0.1728} = 0.421$$
$$-\log(0.421) = 0.375$$
$$\log(1 + W) = \log(1.1) = 0.0413$$

Hence,

$$N = \frac{0.375}{0.0413} = 9.059$$

The payback period of the system is 9.06 years.

13.5.2.3 Alternate Expression for Payback Period, *N* Calculation

$$N = \frac{\log\left[\left\{\left(\frac{I}{C_0 S}\right)\left(\frac{d-1}{d}\right)\right\} + 1\right]}{\log d}$$

Here,

$$d = \frac{1 + e}{1 + r}$$
$$d = \frac{1 + 0}{1 + 0.1} = \frac{1}{1.1}$$

or,

$$\frac{d - 1}{d} = -0.1$$
$$I = 70\,\text{lacs}$$
$$C_0 S = 4.2 \times 100 \times 360 \times 8 = 1209,600$$
$$N = \frac{\log\left[\left\{\left(\frac{7000000}{1209600}\right)\ (-0.1)\right\} + 1\right]}{\log\frac{1}{1.1}} = \frac{\log[-0.578 + 1]}{\log(0.90909)} = \frac{-0.3754}{-0.04139}$$

or,

$$N = 9.069 \text{ years}$$

Table 13.2 The variation of the payback period, *N* in years, for different daily insolation at discount rate of 8%, 10% and 12%

Insolation/day	$r = 8\%$	$r = 10\%$	$r = 12\%$
4.2 kWh	7.28	7.99	8.96
4.8 kWh	6.19	6.70	7.36
5.4 kWh	5.39	5.77	6.23

The variation of the payback period, *N* in years for the daily insolation as 4.2 kWh, 4.8 kWh and 5.4 kWh per day at discount rate of 8%, 10% and 12% with 2% escalation rate of energy cost is illustrated in Table 13.2. These values are the characteristics of Indian region.

References

Adams, W.G. and Day, R.E. (1877). The action of light on selenium. *Proc. R. Soc. London, ser. A*, **25:** 113.

Arab, A.H., Chenlo, F. and Benghanem, M. (2004). Loss of load probability of photovoltaic water pumping system. *Solar Energy*, **76:** 713.

Arthur, D. Little Inc. (1995). Building-Integrated Photovoltaics (BIPV)—Analysis and US Market Potential. Acorn Park, Cambridge, Massachusetts.

Backus, C.E. (1977). Photovoltaic conversion. *In:* Solar Energy Engineering. Academic Press, New York.

Barbaro, S., Coolino, S., Leone, C. and Sinagra, E. (1978). Global solar radiation in Italy. *Solar Energy*, **20:** 431.

Becquerel, E. (1839). On electric effects under the influence of solar radiation. *Comt. Rend.*, **9:** 561.

Bhattacharya, A.B. and Gangadhar, R.B. (1971). Admittance characteristics of a *p-n* junction diode with narrow diffused region, *Solid-State Electron*, **14:** 1113–1118.

Bhattacharya, T. (1998). Terrestrial Solar Photovoltaic. Narosa Publishing House, New Delhi.

Billinton, R. and Allan, R.N. (1992). Reliability evaluation of engineering systems: Concepts and techniques. Plenum Press, New York, USA.

Bird, R.E., Hulstrom, R.L. and Lewis, L.J. (1983). Terrestrial solar spectral data sets. *Solar Energy*, **30(6):** 563–573.

Bishop, J.W. (1988). Computer simulation of the effects of electrical mismatches in photovoltaic cell interconnection circuits. *Solar Cells*, **25:** 75.

Blakers, A.W. and Green M.A. (1986). 20-percent efficient solar cell. *Appl Physics Lett*, **48:** 215–217.

Blakers, A.W., Wang, A., Milne, A.M., Jhao, J. and Green, M.A. (1989). 22.8-percent efficient silicon solar cell. *Appl Physics Letter*, **55(13):** 1363–1365.

Boes, E. (1981). Fundamental of solar radiation. *In:* Solar Energy Handbook. J.F. Kreider and F. Kreith (eds). McGraw Hill Book Company, New York.

Borowy, B.S. and Salameh, Z.M. (1994). Optimum photovoltaic array size for a hybrid wind/PV system. *IEEE Transaction on Energy Conversion*, **9(3):** 482.

Borowy, B.S. and Salameh, Z.M. (1996). Methodology for optimally sizing the combination of a battery bank and a PV array in a wind/PV hybrid system. *IEEE Transaction on Energy Conversion*, **11(2):** 367.

Brandstetter, A. (1984). Solar cell arrays studies. *Proceedings of the Solar World Congress*, Australia, Permagon Press, **4:** 1549.

Brian O' Regan and Gratzel, M. (1991). A low cost high efficiency solar cell based on dye-sensitized colloidal TiO_2 films. *Nature*, **353:** 737–740.

N. D. Kaushika et al., *Solar Photovoltaics*,
https://doi.org/10.1007/978-3-319-72404-1

Bucciarelli, Jr. L.L. (1979). Power loss in photovoltaic array due to mismatch in cell characteristics. *Solar Energy*, **23:** 277.

Bucciarelli, Jr. L.L. (1984). Estimating loss of power probabilities of stand-alone photovoltaic solar energy systems. *Solar Energy*, **32:** 205.

Bucciarelli, Jr. L.L. (1986). The effect of day-to-day correction in solar radiation on the probability of loss of power in standalone photovoltaic solar energy systems. *Solar Energy*. **36(1):** 11.

Casacca, M.A. and Salameh, Z.M. (1992). Determination of Lead-Acid Battery via Mathematical Modeling Techniques. *IEEE Transaction on Energy Conversion*, **7(3):** 442.

Chamberlin, C.E., Lehman, P., Zoellick, J. and Pauletto, G. (1995). Effect of mismatch losses in photovoltaic arrays. *Solar Energy*, **54(5):** 165.

Chambouleyron, I. (1991). Light induced meta-stability in a-Ge:H. *Journal of Non-Crystal Solids*, **128(3):** 243–254.

Chaplin, D.M., Fuller, C.S. and Pearson, G.L. (1954). A new silicon p-n junction photocell for converting solar radiation into electrical power. *Journal of Applied Physics*, **25:** 676–677.

Chedid, R., Akiki, H. and Rahman, S. (1998). A decision support technique for the design of hybrid solar-wind power systems. *IEEE Transaction on Energy Conversion*, **13(1):** 76.

Chemmangot, V.N., Ashari, M. and Keerthipala, W.W.L. (2000). A Grid-Interactive Uninterruptible Power Supply System Using Battery Storage and a Back up Diesel Generator. *IEEE Transaction on Energy Conversion*, **15(3):** 348.

Chenni, R., Makhlour, M., Kerbache, T. and Bouzied, A. (2007). A detailed modeling method for photovoltaic cell. *Energy*, **32(9):** 1724–1730.

Chowdhary, B.H. (1992). Optimizing the integration of photovoltaic systems with electric utilities. *IEEE Transaction on Energy Conversion*, **7(1):** 72.

Chowdhary, B.H. and Rahman, S. (1988). Analysis of Interrelationship between photovoltaic power and battery storage for Electric Utility Load Management. *IEEE Transaction on Energy Conversion*, **3(3):** 900.

Dorvlo, A.S.S., Jervase, J.A. and Al-Lawati, A. (2002). Solar radiation assessment using artificial neural network. *Applied Energy*, **71:** 307–319.

Duffie, J.A. and Beckman, W.A. (1980). Solar Engineering Processes. John Wiley and Sons, New York.

El-Nashar, A.M. (1981). Solar radiation characteristics in Abu Dhabi. *Solar Energy*, **47(1):** 49.

Endo, E., Kurokawa, K. and Hyaashi, M. (1992). Design aid expert system for photovoltaic power generation system. 2nd Renewable Energy Congress, Reading England, Sept. 13–18.

Enslin, J., Wolf, M., Snyman, D. and Swiegrs, W. (1997). Integrated photovoltaic maximum power tracking converter. *IEEE Trans. Ind. Electron*, **44(6):** 769.

Esram, T. and Chapman, P.L. (2007). Comparison of Photovoltaic Array Maximum Power Point Tracking Techniques. *IEEE Transaction on Energy Conversion*, **22(2):** 439–449.

Fahrenbruch, A.L. and Bube, R.H. (1983). Application of the transport equation in fundamentals of solar cells. Academic Press, New York.

Fam, W.Z. and Balachander, M.K. (1988). Dynamic performance of a dc shunt motor connected to photovoltaic array. *IEEE Transactions on Energy Conversion*, **3:** 613.

Feldman, J., Singer, S. and Braunstein, A. (1981). Solar cell interconnections and the shadow problem. *Solar Energy*, **26:** 419–428.

Fraunhofer, I.S.E. (2016). Photovoltaics Report, 17 November 2016 (www.ise.fraunhofer.de).

Francois, G. and Salameh, Z.M. (2001). Steady-State Performance of a Grid-Connected Rooftop Hybrid Wind-Photovoltaic Power System with Battery Storage. *IEEE Transaction on Energy Conversion*, **16(1):** 1.

Gangadhar, R.B., Mathur, R.S., Sehgal, H.K. and Kulshreshta, A.P. (1967). Investigations of silicon p-n junction for solar energy conversion. *Indian Journal of Pure and Applied Physics*, **5:** 593–595.

Gangadhar, R.B. (1997). Investigation of silicon solar cells and their power supply systems. Ph.D. Thesis (Supervisor N.D. Kaushika).

Gangadhar, R.B. and Bhattacharya, A.B. (1968). Drift field photovoltaic cell performance with bulk and surface recombination. *International Journal of Electron*, **25:** 17–26.

Gangadhar, R.B. and Bhattacharya, A.B. (1969). On the transport characteristics of a p-n junction with narrow diffused region. *Solid State Electronics*, **12:** 887–891.

Gangadhar, R.B. and Kaushika, N.D. (1990). Solar spectral response of graded base silicon cell. Proc. National Energy Solar Convention, Udaipur, pp. 397–401.

Gangadhar, R.B. and Kaushika, N.D. (1992a). Physics and Technology of Solar Cells. *In:* Energy Resources and Technology. N.D. Kaushika, S. Kaneff and C.S. Jha (eds). Scientific Publishers, Jodhpur.

Gangadhar, R.B. and Kaushika, N.D. (1992b). Spectral response of graded base solar cells. *Energy Conversion Management*, **33(4):** 275–278.

Gao, F., Wang, Y., Zhang, J. Shi, D., Wang, M., Humphry-baker, R., Wang, P. and Zakeeruddin, M. Gratezel (2008). A new heteroleptic ruthenium sensitizer enhances the absorptivity of mesoporous titania film for a high efficiency dye-sensitized solar cell. *Chemical Communications*, **23:** 2635–2637.

Gautam, N.K. and Kaushika, N.D. (2001). Network analysis of fault-tolerant solar photovoltaic arrays. *Solar Energy Material and Solar Cell*, **69:** 25.

Gautam, N.K. and Kaushika, N.D. (2002a). An efficient algorithm to simulate the electrical performance of solar photovoltaic arrays. *Energy*, **27:** 347.

Gautam, N.K. and Kaushika, N.D. (2002b). Reliability evaluation of solar photovoltaic arrays. *Solar Energy*, **72(2):** 129.

Gavanidou, E.S. and Bakirtzis, A.G. (1992). Design of stand-alone system with renewable energy sources using trade-off methods. *IEEE Transaction on Energy Conversion*, **7(1):** 42.

Glasnovic, Z. and Margeta, J. (2006). A model for optimum sizing of photovoltaic irrigation water pumping system. *Solar Energy*, **81(7):** 904–916.

Goh, T.N. (1979). Statistical study of solar radiation in formation in an equatorial region (Singapore). *Solar Energy*, **22,** 105.

Gonzalez, C. and Weaver, R. (1980). Circuit design considerations for photovoltaic modules and systems. Proceedings of the 14th IEEE Photovoltaic Specialists Conferences, San Diego, New York, USA.

Gover, A. and Stella, P. (1974). Vertical multi-junction solar-cell one-dimensional analysis. *IEEE Transaction on Electron Devices*, **ED-21:** 351.

Gow, J. and Manning, C. (2000). Controller management for boost converter systems source from solar photovoltaic arrays or other maximum power sources. *IEE Proc. Electronics Power Appl.*, **147(1):** 15.

Green, M.A. (1979). Solar cells—Operating principle, technology and system applications. A short course on solar photovoltaic systems at University of Queensland, Aug. 21–23.

Green, M.A., Gauja, E. and Withayachamnankul, W. (1981). Silicon solar cells with integral bypass diode. *Solar Cells*, **3:** 233–244.

Green, M.A. (1982). Solar cells. Prentice-Hall Inc., Englewood Cliffs, NJ.

Green, M.A. (1984), Limits on the open circuit voltage and efficiency of silicon solar cells imposed by intrinsic auger processes. *IEEE Trans Electron Devices*, **ED-31:** 671–678.

Green, M.A. (1987). High efficiency silicon solar cells. Aedermaunsdorf: Trans. Tech. Publications.

Green, M.A., Blaker, A.W., Zhao, J., Milne, A.M., Wang, A. and Dai, X. (1990). Characterization of 23 percent efficient silicon solar cells. *IEEE Trans. on Electron Devices*, **37(2):** 331–335.

Green, M.A. et al. (2011). Solar Cell Efficiency Tables (Version 37). Progress in Photovoltaics: Research and Applications, Vol. 19. John Wiley & Sons Ltd., N.J.

Grondahl, L.O. (1933). The copper-cuprous oxide rectifier and electric cell. *Rev. Mod Physics*, **5:** 141.

Gunes, Serap., Neugebaucer, H. and Sariciftci, N.S. (2007). Conjugated polymer based organic solar cells. *Chem. Rev.*, **107:** 1324–1338.

Halls, J.J.M., Friend, R.H., Archer, M.D. and Hill, R.D. (2001). Clean electricity from photovoltaics. Imperial College Press, London.

Hilloowala, R.M. and Sharaf, A.M. (1992). A Rule-Based Fuzzy Logic Controller for a PWM Inverter in Photovoltaic Energy Conversion Scheme. Proc. IEEE IAS Conf., Houston.

Hontoria, L., Anguilera, J. and Zufiria, P. (2005). A new approach for sizing stand alone photovoltaic systems based in neural network. *Solar Energy*, **78(2):** 313.

Hoppe, H. and Sariciftci, N.S. (2004). Organic solar cells: An overview. *J. Mater. Res.*, **19:** 1924–1945.

Hottel, H.C. (1976). A simple model for estimating the transmittance of direct solar radiation through clear atmosphere. *Solar Energy*, **18:** 129.

Hussein, H.M.S., Ahmad, G.E. and El-Ghetany, H.H. (2004). Performance evaluation of photovoltaic modules at different tilt angles and orientation. *Energy Conversion and Management*, **45:** 2441–2452.

Jackson, E. (1955). Areas for improvement of the semiconductor solar energy converter. *In:* Transactions of the conference on the use of solar energy, Vol. 5.

Jain, G.C. and Al-Rifai, R.M.S. (1967). Effects of Drift Fields and Field Gradients on the Quantum Efficiency of Photocells. *Journal of Applied Physics*, **38:** 768.

Jain, S.C., Sehgal, H.K. and Gangadhar, R.B. (1968). Silicon Solar Cells and Their Terrestrial Applicatins. *Journal of Scientific & Industrial Research*, **27(8):** 295–305.

Jordan, A.G. and Milnes, A.G. (1960). Photoeffect on diffused P-N junctions with integral field gradients. *I R E Transaction on Electron Devices*, **ED-7 (4):** 242–251.

Kalogivou, S.A. and Tripanagnostopoulos, Y. (2006). Hybrid PV/T systems for domestic hot water and electricity production. *Energy Conversion and Management*, **47(18–19):** 3368.

Kaneff, S. (1983). On the practical and potential viability of paraboloidal dish solar thermal electric power systems. Proc. Solar World Congress (ISES) Perth, 4, p. 2767.

Karatepe, E., Mutlu, B. and Colak, M. (2006). Neural network based solar cell model. *Energy Conversion and Management*, **47(9–10):** 1159.

Karatepe, E., Mutlu, B. and Colak, M. (2007). Development of suitable model for characterizing photovoltaic arrays with shaded solar cell. *Solar Energy*, **81:** 977–992.

Kaushika, N.D. (1986). Solar parabolic dish power production for rural electrification. *Invention Intelligence*, May-June, 182.

Kaushika, N.D. (1996). 50 kW solar PV power plant in Rajasthan TEFR. Sun Source of India Ltd.

Kaushika, N.D. (1999). Design and development of fault tolerant circuitry to improve the reliability of solar PV modules and arrays. Final Technical report of the Department of Science and Technology, the Government of India Research project no. III 5 (98) / 95-ET.

Kaushika, N.D. and Gautam, N.K. (2001). Mismatch losses and time to failure of solar PV arrays. Proceedings International Solar Energy Society Meeting, Adelaide, Australia.

Kaushika, N.D. (2002). Solar PV design aid expert system. *In:* Final Technical report of the Department of Science and Technology, the Government of India Research project no. III 5 (81) / 99-ET.

Kaushika, N.D. and Gautam, N.K. (2003). Energy Yield Simulations of Interconnected solar PV Arrays. *IEEE Transaction on Energy Conversion*, **18(1):** 127.

Kaushika, N.D., Gautam, N.K. and Kaushik, K. (2005). Simulation model for sizing of standalone solar PV system with interconnected array. *Solar Energy Materials and Solar Cell*, **85(4):** 499.

Kaushika, N.D. and Rai, Anil K. (2006). Solar PV design aid expert system. *Solar Energy Materials and Solar Cells*, **90:** 2829–2845.

Kaushika, N.D. and Rai, Anil K. (2007). An investigation of mismatch losses in solar Photovoltaic cell networks. *Energy*, **32:** 755–759.

Kaushika, N.D., Tomar, R.K. and Kaushik, S.C. (2014). Artificial neural network model based on interrelationship of direct, diffuse and global solar radiations. *Solar Energy*, **103:** 327–342.

Kazmerski, L.L. and Emery, K.A. (1992). Proc. 2nd World Renewable Energy Congress, Reading, U.K.

Kellogg, W.D., Nehrir, M.H., Venkataramanan, G. and Gerez, V. (1998). Generation unit sizing and cost analysis for stand-alone wind, photovoltaic and hybrid wind/PV systems. *IEEE Transaction on Energy Conversion*, **13(1):** 70.

Khouzem, K. and Khouzam, L. (1993). Optimum matching of Direct coupled Electromechanical Loads to a photovoltaic generator. *IEEE Transaction on Energy Conversion*, **8(3):** 343.

Kislovski, A. and Redl, R. (1994). Maximum power tracking using positive feedback. *Proc. IEEE Power Electronics Spec. Conf.* 1065.

Klein, S.A. and Beckman, W.A. (1987). Loss-of-load probabilities for standalone photovoltaic systems. *Solar Energy*, **39(6):** 499.

Konadu, R.A. and Mingyua, C. (1999). An Expert System tool for designing hybrid energy systems. *Power and Energy Systems*, **19(3):** 239.

Koutroulis, E., Kalaitzakis, K. and Voulgaris, N.C. (2001). Development of a micro controller based photovoltaic maximum power point tracking control system. *IEEE Transaction Power Electron*, **16(21):** 46–54.

Koutroulis, E., Kolokotsa, D., Potivakis, A. and Kalaitzakis, K. (2006). Methodology for optimal sizing of stand-alone photovoltaic/wind generation system using genetic algorithm. *Solar Energy*, **80(9):** 1072.

Landsman, E.E. (1978). Maximum power trackers for photovoltaic array. 13th IEEE PVSC, Washington DC, June 5–8.

Lange, B. (1930). New photoelectric cell. *Zeit Physics*, **31:** 139.

Lee, Tsung-Ying and Chen, N. (1993). Optimal capacity of the battery energy storage system in a power system. *IEEE Transaction on Energy Conversion*, **8(4):** 667.

Liu, B.Y.H. and Jordan, R.C. (1960). The Interrelationship and Characteristic Distribution of Direct, Diffuse and Total Solar Radiation. *Solar Energy*, **4:** 1–19.

Liu, B.Y.H. and Jordan, R.C. (1962). Daily Insolation on Surfaces Tilted towards the Equator. *ASHRAE Transactions*, **67:** 526–541.

Lokhande, C.D. and Dhawla, D.S. (2009). Chemically deposited thin film for solar cells. Proc. National Conference on Renewable Energy (NACORE-2009), Jodhpur, India.

Mack, M.R. (1979). Solar Power for Telecommunication. *Telecommunication Journal of Australia*, **29(1):** 120.

Mani, A. (1980). Handbook of Solar Radiation Data for India. Allied Publishers, India.

Marwali, M.K.C., Shahidetipour, S.M. and Daneshdoot, M. (1997). Probabilistics-utility systems with battery storage. *IEEE Transaction on Energy Conversion*, **12(2):** 175.

McDaniels, D.K. (1979). The Sun: Our future energy source. John Wiley and Son Inc., New York.

Mikroyannidis, John A., Stylianakis, Minas M., Suresh, P. and Sharma, G.D. (2009). Efficient hybrid bulk heterojunction solar cells based on phenylenevinylene copolymer, perylene bisimide and TiO_2. *Solar Energy Material & Solar Cells*, **93:** 1792–1800.

Milstein, J.B. and Tsno, Y.S. (1984). Research on crystalline silicon solar cells. Proc. 17th IEEE Photovolt. Special Conf. Orlando, FL.

Mohan, N. (2004). Power Electronics Converters, Application and Design. Tore M. Undeland and William P. Robbins (eds). John Wiley & Sons (Asia).

Mondol, J.D., Yohanis, Y.G. and Norton, B. (2006). Optimal sizing of array and inverter for grid connected photovoltaic systems. *Solar Energy*, **80(12):** 1517.

Nozik, A.J. (2001). Spectroscopy and hot election relaxation dynamics in semiconductor quantum wells and quantum dots. *Annual Reviews of Physics and Chemistry*, **52,** 193.

Ofry, E. and Braunstein, A. (1983). The Loss of Power Supply Probability as a Technique for Designing Stand-alone Solar Photovoltaic systems. *IEEE Transaction on Power Apparatus & Systems*, **102(5):** 1171.

Ogeman, H., Ecevit, A. and Tasdemiroglu, E. (1984). A new method for estimating solar radiation from bright sunshine data. *Solar Energy*, **19:** 307.

Pulfrey, L.D. (1978). Photovoltaics Power Generation. Van Nostrand Reinhold Co., New York.

Quaschning, V. and Hanitsch, R. (1996). Numerical simulation of current-voltage characteristics of photovoltaic systems with shaded solar cells. *Solar Energy*, **56:** 513.

Rai, Anil K., Kaushika, N.D., Singh, B. and Agarwal, N. (2011). Simulation model of ANN based maximum power point tracking controller for solar PV system. *Solar Energy Materials & Solar Cells*, **95:** 773–778.

Ralph, E.L. (1965). Use of concentrated sunlight with solar cells for terrestrial uses. Presented at the Solar Energy Society Annual Meeting, Arizona.

Ramkumar, R., Shetty, P.S. and Ashenayi, K. (1986). A Linear Programming Approach to the Design of Integrated Renewable Energy Systems for Developing Countries. *IEEE Transaction on Energy Conversion*, **7 EC-1(4):** 18.

Ramkumar, R., Abouzahr, I. and Ashenayi, K. (1992). A knowledge-based approach to the design of Integrated Renewable Energy Systems. *IEEE Transaction on Energy Conversion*, **7(4):** 648.

Raushenbach, S.H. (1971). Electrical output of shadowed solar cell arrays. *IEEE Transaction on Electron Devices*, **18:** 483.

Ravich, Yu. L. (1962). Radiotekh. *Elektron*, **7:** 1002.

Ray, S. (2009). Nanocrystalline silicon based thin film solar cell. Proc National Conference on Renewable Energy (NACORE-2009), Jodhpur, India.

Reddy, S.J. (1971). An empirical method for estimation of the total solar radiation. *Solar Energy*, **14:** 289.

Reynolds, D.C., Leies, G., Antes, L.L. and Marburger, R.E. (1954). Photoelectric effect in cadmium sulphide. *Physics Rev.*, **96:** 533–534.

Riad, C. and Rahman, S. (1997). Unit sizing and control of hybrid wind-solar power systems. *IEEE Transaction on Energy Conversion*, **12(1):** 79.

Robbins, T.P. (1987). Simulation of a Stand-alone Power System with Battery Storage. Annual Conference and Annual General Meeting of Anzses, Australian National University, 27.

Ross, R.G. (1981). Photovoltaic module and array reliability. Proc. of 15th IEEE Photovoltaic Specialists Conf., Orlando, FL, New York, USA.

Runyan, W.R. (1965). Silicon Semiconductor technology. McGraw Hill, NewYork.

Runyan, W.R. and Alexander, E.G. (1967). *IEEE Trans. Electron Devices*, **14:** 3.

Salameh, Z. and Taylor, D. (1990). Step-up maximum power point tracker for photovoltaic arrays. *Solar Energy*, **44:** 57–61.

Sabbagh, J.A., Sayigh, A.A.M. and El-Salam, E.M.A. (1977). Estimation of the total solar radiation from meteorological data. *Solar Energy*, **68(2):** 169–178.

Salameh, Z.M., Borowy, B.S. and Amin, A.R.A. (1995). Photovoltaic module-site matching based on the capacity factors. *IEEE Transaction on Energy Conversion*, **10(2):** 326.

Sayed, M. and Partin, L. (1975). Effect of shading on CdS/Cu_xS solar cell array design. *IEEE Trans Energy Conversion*, **14:** 61.

Schottky, W. (1930). Cuprous oxide photoelectric cells. *Zeit Physics*, **31:** 913.

Sfetsos, A. and Coonick, A.H. 2000. Univariate and multivariate forecasting of hourly solar radiation with artificial intelligence techniques. *Solar Energy*, **68(2):** 169–178.

Shah, A.V., Meir, J., Vallat-Sauvair, E., Wyrsch, N., Kroll, U., Droz, C. and Graf, U. (2003). Material and solar cell research in microcrystalline silicon. *Solar Energy Material & Solar Cells*, **78:** 469–491.

Sharma, G.D., Suresh, P., Mikroyannidis, John A. and Stylianakis, Minas M. (2010). Organic PV devices based on conjugate urinyle copolymer: PCBM bled with incorporation of low band gas small molicunk. *J. Mater. Chem*, **20:** 561–567.

Sharma, G.D. (2009). Advances in Nano Structured Organic Solar Cells. Proc National Conference on Renewable Energy (NACORE-2009), Jodhpur, India.

Shephard, N.F. and Sugimura, R.S. (1984). The Integration of bypass diode with Terrestrial Photovoltaic Modules and Arrays. Proc. of the 17th IEEE Photovoltaic Specialists Conf.

Shrestha, G.B. and Goel, L. (1998). A study on optimal sizing of stand-alone photovoltaic stations. *IEEE Transaction on Energy Conversion*, **13**(4): 373.

Singer, S., Rosenshtein, B. and Surazi, S. (1984). Characterization of PV array output using a small number of measured parameter. *Solar Energy*, **32:** 603.

Siva Kumar, R. and Anandan, S. (2009). New type of inorganic-organic Hybrid (Heteropoly tungstic acid-Polyepich lorohydrin) Polymer electrolyte with TiO_2 nano filler for solid state dye sensitized solar cells. Proc National Conference on Renewable Energy (NACORE-2009), Jodhpur, India.

Smith, K.D., Gummel, H.K., Bode, J.D., Cuttriss, D.B., Nielsen, R.J. and Rosenzweig, W. (1963). The Solar Cells and Their Mounting. *Bell System Technical Journal*, **42:** 1765–1816.

Solar Directory (1989). Prospectus for Renewable Energy in Australia. ANZSES Report.

Stember, L.H. (1981). Reliability considerations in the design of solar photovoltaic power system. *Solar Cells*, **3:** 269–285.

Sukhatme, S.P. (1984). Solar Energy Principles of Thermal Collection and Storage. Tata McGraw-Hill, New Delhi.

Supit, I. and Van Kappel, R.R. (1998). A simple method to estimate global radiation. *Solar Energy*, **63:** 147.

Suresh, P., Sharma, G.D., Mikroyannidis, John A. and Stylianakis, Minas M. (2009). Organic Photovoltaic devices based on conjugated rinyle copolymer: PCBM blend with incorporation of low band gap small molecule. Proc National Conference on Renewable Energy (NACORE-2009), Jodhpur, India.

Swalesh, M.S. and Green, M.A. (1982). Effect of shunt resistance and bypass diode on the shadow tolerance of solar cell modules. *Solar Cells*, **5:** 183.

Theraja, B.L. and Theraja, A.K. (1988). A text-book of electrical technology. Nirja Construction & Development Co. (P) Ltd, New Delhi.

Tiedje, T., Vablonowitch, E., Cody, G.D. and Brooks, B.G. (1984). Limiting efficiency of silicon solar cells. *IEEE Trans. on Electron Devices*, **ED-31:** 711–761.

Treble, F.C. (1991). Generating electricity from Sun. Pergamon Press, New York.

Veerachary, M., Senjyu, T. and Uezeto, K. (2002a). Feed forward maximum power point tracking of PV systems using fuzzy controller. *IEEE Transactions on Aerospace and Electronic System*, **38(3):** 969.

Veerachary, M., Senjyu, T. and Uezeto, K. (2002b). Voltage based maximum power point tracking of PV systems. *IEEE Transactions on Aerospace and Electronic System*, **38(1):** 262.

Wang, A., Zhao, J. and Green, M.A. (1990). 24-percent efficient silicon solar cells. *Appl. Phys. Lett.*, **57:** 602–604.

Wasynczuk, O. (1989). Modeling and Dynamic Performance of a Line-Commutated Photovoltaic Inverter System. *IEEE Transaction on E.C.*, **4(3):** 37.

Winter, C.J., Sizmann, L.L. and Vant-Hull (1991). Solar Power Plants. Springer-Verlag Publishers.

Wolf, H.F. (1969). Silicon semiconductor data. Pergamon Press, New York.

Wolf, M. (1960). Limitations and possibilities for improvement of photovoltaic energy converters, Part I: Considerations for Earth's surface operation. *Proc. Inst Radio Engg*, **48:** 1246–1263.

Wolf, M. (1963). Drift fields in photovoltaic solar energy converter cell. *Proceedings IEEE*, **5:** 674.

Wolf, M. (1971). A new look at silicon solar cell performance. *Energy Conversion*, **11:** 63–73.

Wolf, M. and Ralph, E.L. (1965). Effects of thickness on short-circuit current of silicon solar cells. *IEEE Transactions on Electron Devices*, **12(8):** 470.

Wolf, S. and Enslin, J. (1993). Economical PV maximum power point tracking regulator with simplistic controller. *Proc. IEEE Power Electronics Spec. Conf.* 581.

Yella, A., Lee, H.W., Tsao, H.N., Yi, C., Chandiran, A.K., Nazeeruddin, M.K., Diau, EW-D., Yeh, C.Y., Zakeeruddin, S.M. and Gratzel, M. (2011). Porphyrin-sensitized solar cells with cobalt (II/III)-based redox electrolyte exceed 12 percent efficiency. *Science*, **6056:** 629–634.

Yokoyama, R., Ito, K. and Yuasa, Y. (1994). Multiobjective optimal unit sizing of hybrid power generation systems utilizing photovoltaic and wind energy. *International Journal of Solar Energy Engineering*, **116:** 167.

Index

N. D. Kaushika et al., *Solar Photovoltaics*,
https://doi.org/10.1007/978-3-319-72404-1

MIX
Papier aus verantwortungsvollen Quellen
Paper from responsible sources
FSC® C105338

If you have any concerns about our products,
you can contact us on
ProductSafety@springernature.com

In case Publisher is established outside the EU,
the EU authorized representative is:
Springer Nature Customer Service Center GmbH
Europaplatz 3, 69115 Heidelberg, Germany

Printed by Libri Plureos GmbH
in Hamburg, Germany